Ceramics in
Nuclear Applications

T0324851

Ceramics in Nuclear Applications

*A Collection of Papers Presented at the
33rd International Conference on
Advanced Ceramics and Composites
January 18–23, 2009
Daytona Beach, Florida*

Edited by
**Yutai Katoh
Alex Cozzi**

Volume Editors
**Dileep Singh
Jonathan Salem**

A John Wiley & Sons, Inc., Publication

Copyright © 2010 by The American Ceramic Society. All rights reserved.

Published by John Wiley & Sons, Inc., Hoboken, New Jersey.
Published simultaneously in Canada.

No part of this publication may be reproduced, stored in a retrieval system, or transmitted in any form or by any means, electronic, mechanical, photocopying, recording, scanning, or otherwise, except as permitted under Section 107 or 108 of the 1976 United States Copyright Act, without either the prior written permission of the Publisher, or authorization through payment of the appropriate per-copy fee to the Copyright Clearance Center, Inc., 222 Rosewood Drive, Danvers, MA 01923, (978) 750-8400, fax (978) 750-4470, or on the web at www.copyright.com. Requests to the Publisher for permission should be addressed to the Permissions Department, John Wiley & Sons, Inc., 111 River Street, Hoboken, NJ 07030, (201) 748-6011, fax (201) 748-6008, or online at http://www.wiley.com/go/permission.

Limit of Liability/Disclaimer of Warranty: While the publisher and author have used their best efforts in preparing this book, they make no representations or warranties with respect to the accuracy or completeness of the contents of this book and specifically disclaim any implied warranties of merchantability or fitness for a particular purpose. No warranty may be created or extended by sales representatives or written sales materials. The advice and strategies contained herein may not be suitable for your situation. You should consult with a professional where appropriate. Neither the publisher nor author shall be liable for any loss of profit or any other commercial damages, including but not limited to special, incidental, consequential, or other damages.

For general information on our other products and services or for technical support, please contact our Customer Care Department within the United States at (800) 762-2974, outside the United States at (317) 572-3993 or fax (317) 572-4002.

Wiley also publishes its books in a variety of electronic formats. Some content that appears in print may not be available in electronic format. For information about Wiley products, visit our web site at www.wiley.com.

Library of Congress Cataloging-in-Publication Data is available.

ISBN 978-0-470-45760-3

10 9 8 7 6 5 4 3 2 1

Contents

MATERIAL AND COMPONENT PROCESSING

CERAMICS FOR FUEL COATING

NUCLEAR FUELS AND WASTES

Preface

Presentations on Ceramics in Nuclear Applications took place in conjunction with the 33rd International Conference and Exposition on Advanced Ceramics and Composites, January 18–23, 2009 in Daytona Beach, Florida. This volume documents a number of the papers presented during the International Symposium on Silicon Carbide and Carbon-Based Materials for Fusion and Advanced Nuclear Energy Applications and the focused session on Processing and Properties of Nuclear Fuels and Wastes. The symposium and focused session were sponsored by The American Ceramic Society's (ACerS) Engineering Ceramics and Nuclear and Environmental Technology Divisions respectively.

The success of the sessions and the issuance of the proceedings could not have been possible without the support of ACerS staff and the other organizers of the program. Their assistance, along with that of the session chair's, was invaluable in ensuring the creation of quality proceedings.

YUTAI KATOH
ALEX COZZI

Introduction

The theme of international participation continued at the 33rd International Conference on Advanced Ceramics and Composites (ICACC), with over 1000 attendees from 39 countries. China has become a more significant participant in the program with 15 contributed papers and the presentation of the 2009 Engineering Ceramic Division's Bridge Building Award lecture. The 2009 meeting was organized in conjunction with the Electronics Division and the Nuclear and Environmental Technology Division.

Energy related themes were a mainstay, with symposia on nuclear energy, solid oxide fuel cells, materials for thermal-to-electric energy conversion, and thermal barrier coatings participating along with the traditional themes of armor, mechanical properties, and porous ceramics. Newer themes included nano-structured materials, advanced manufacturing, and bioceramics. Once again the conference included topics ranging from ceramic nanomaterials to structural reliability of ceramic components, demonstrating the linkage between materials science developments at the atomic level and macro-level structural applications. Symposium on Nanostructured Materials and Nanocomposites was held in honor of Prof. Koichi Niihara and recognized the significant contributions made by him. The conference was organized into the following symposia and focused sessions:

Symposium 1	Mechanical Behavior and Performance of Ceramics and Composites
Symposium 2	Advanced Ceramic Coatings for Structural, Environmental, and Functional Applications
Symposium 3	6th International Symposium on Solid Oxide Fuel Cells (SOFC): Materials, Science, and Technology
Symposium 4	Armor Ceramics
Symposium 5	Next Generation Bioceramics
Symposium 6	Key Materials and Technologies for Efficient Direct Thermal-to-Electrical Conversion
Symposium 7	3rd International Symposium on Nanostructured Materials and Nanocomposites: In Honor of Professor Koichi Niihara
Symposium 8	3rd International symposium on Advanced Processing & Manufacturing Technologies (APMT) for Structural & Multifunctional Materials and Systems

Symposium 9	Porous Ceramics: Novel Developments and Applications
Symposium 10	International Symposium on Silicon Carbide and Carbon-Based Materials for Fusion and Advanced Nuclear Energy Applications
Symposium 11	Symposium on Advanced Dielectrics, Piezoelectric, Ferroelectric, and Multiferroic Materials
Focused Session 1	Geopolymers and other Inorganic Polymers
Focused Session 2	Materials for Solid State Lighting
Focused Session 3	Advanced Sensor Technology for High-Temperature Applications
Focused Session 4	Processing and Properties of Nuclear Fuels and Wastes

The conference proceedings compiles peer reviewed papers from the above symposia and focused sessions into 9 issues of the 2009 Ceramic Engineering & Science Proceedings (CESP); Volume 30, Issues 2-10, 2009 as outlined below:

- Mechanical Properties and Performance of Engineering Ceramics and Composites IV, CESP Volume 30, Issue 2 (includes papers from Symp. 1 and FS 1)
- Advanced Ceramic Coatings and Interfaces IV Volume 30, Issue 3 (includes papers from Symp. 2)
- Advances in Solid Oxide Fuel Cells V, CESP Volume 30, Issue 4 (includes papers from Symp. 3)
- Advances in Ceramic Armor V, CESP Volume 30, Issue 5 (includes papers from Symp. 4)
- Advances in Bioceramics and Porous Ceramics II, CESP Volume 30, Issue 6 (includes papers from Symp. 5 and Symp. 9)
- Nanostructured Materials and Nanotechnology III, CESP Volume 30, Issue 7 (includes papers from Symp. 7)
- Advanced Processing and Manufacturing Technologies for Structural and Multifunctional Materials III, CESP Volume 30, Issue 8 (includes papers from Symp. 8)
- Advances in Electronic Ceramics II, CESP Volume 30, Issue 9 (includes papers from Symp. 11, Symp. 6, FS 2 and FS 3)
- Ceramics in Nuclear Applications, CESP Volume 30, Issue 10 (includes papers from Symp. 10 and FS 4)

The organization of the Daytona Beach meeting and the publication of these proceedings were possible thanks to the professional staff of The American Ceramic Society (ACerS) and the tireless dedication of the many members of the ACerS Engineering Ceramics, Nuclear & Environmental Technology and Electronics Divisions. We would especially like to express our sincere thanks to the symposia organizers, session chairs, presenters and conference attendees, for their efforts and enthusiastic participation in the vibrant and cutting-edge conference.

DILEEP SINGH and JONATHAN SALEM
Volume Editors

Silicon Carbide
and Carbon Composites

SINGLE- AND MULTI-LAYERED INTERPHASES IN SiC/SiC COMPOSITES EXPOSED TO SEVERE CONDITIONS: AN OVERVIEW

Roger Naslain[*], René Pailler and Jacques Lamon
Laboratory for ThermoStructural Composites (LCTS), 3 Allée de La Boetie,
33600 Pessac, France

ABSTRACT

Pyrocarbon (PyC), which is presently the best interphase material for SiC/SiC composites is not stable under conditions encountered in advanced applications: it is sensitive to oxidizing atmospheres or to neutron irradiation, even at moderate temperatures. Attempts have been pursued to replace PyC-interphase by boron nitride, which is more resistant to oxidation but poorly compatible with nuclear applications. Other interphase materials, such as ternary carbides (MAX-phases) seem promising but their use in SiC/SiC has not been demonstrated. Hence, the most efficient way to improve the behavior of PyC-interphase in severe environments is presently to replace part of the pyrocarbon by a second material displaying a better compatibility with the environment, such as SiC itself ((PyC-SiC)$_n$ multilayered interphases). Important issues related to the design and behavior of layered interphases are reviewed on the basis of LCTS-experience and recent data published by others with a view to demonstrate their potential interest in HT-nuclear reactors.

1- INTRODUCTION

The interphase plays a key role in the behavior of ceramic matrix composites (CMCs). It prevents the early failure of the fibers, matrix microcracks being arrested or/and deflected parallel to fiber axis (so-called "mechanical fuse" function). It also transfers load from the fibers to the matrix and eventually releases part of residual thermal stresses. The interphase protects fibers against chemical reactions that could occur during processing and use of CMCs in aggressive environments.

It has been postulated, that the best interphase materials for SiC/SiC might be those with a *layered structure*, the layers being parallel to fiber surface, weakly bonded to one another but strongly adherent to fibers [1-4]. It appeared that pyrocarbon was the best interphase material for SiC/SiC in terms of their mechanical behavior [1-5].

Unfortunately, pyrocarbon is oxidation prone even at low temperatures with the result that PyC-interphase can be consumed and the FM-coupling lost. Hence, that interphase becomes the weak point of SiC/SiC when used in *oxidizing atmospheres* (gas turbines). Within the same conceptual framework, two alternatives have been proposed: boron nitride (BN) and (X-Y)$_n$ multilayers. The former displays a structure similar to that of graphite while being more oxidation resistant [5-7]. In the latter, with X = PyC or BN, Y = SiC and n = 1 to 10, part of the oxidation prone X-constituent is replaced by a material Y exhibiting a better oxidation resistance, such as SiC itself [1-4, 8-12]. Furthermore, this concept of material multilayering has been extended to the matrix, yielding "self-healing" composites with outstanding lifetimes in oxidizing atmospheres [10, 13, 14].

More recently, SiC/SiC composites have been envisaged as structural materials in high temperature nuclear reactors. The interphase appears again as a possible weak point, PyC being known to undergo volume change when exposed to neutrons whereas BN undergoes nuclear reactions [15-17]. Research is presently pursued following two similar routes: (i) use of thin single PyC-layers and (ii) use of (PyC-SiC)$_n$ multilayers, to minimize the effect of neutron irradiation [15-21].

The aim of the present overview is to recall the basis of the layered interphase concept, to discuss its application to SiC/SiC exposed to oxidizing environment and tentatively, to neutron irradiation.

---

[*] Emeritus

3

2-   PYROCARBON   SINGLE   LAYER INTERPHASE: THE REFERENCE

Pyrocarbon has a structure similar to that of graphite but the elementary graphene layers are of limited size and stacked with rotational disorder. Our layered interphase concept requires that: (i) the layers should be oriented parallel to fiber surface and (ii) the bonding between the fiber and the PyC-interphase should be strong enough. Otherwise, debonding/crack deflection would occur at the fiber surface exposing the fiber to mechanical damage and to the atmosphere [3, 4].

2.1- Pyrocarbon texture

Pyrocarbon displays a variety of microtexture and anisotropy, which can be characterized by optical microscopy in polarized light (extinction angle, A_e) or/and transmission electron microscopy (L_2, N parameters): the larger A_e, the higher the anisotropy [22]. The preferred PyC for an interphase is *rough laminar* ($A_e > 18°$) This RL-PyC has a tendency to grow with graphene layers parallel to fiber surface. It is usually deposited by CVI (I-CVI or P-CVI) from propane or propylene [1, 5, 9, 23].

2.2- Fiber/PyC-interphase bonding

The second requirement is a strong-bonding between the fiber and the PyC-interphase [3, 4]. Achieving such a strong bonding is first a matter of surface chemistry. The surface of desized Si-C-O fibers (Nicalon) is enriched in oxygen and free carbon. As a result, there is in SiC/SiC (CVI) a thin and irregular dual layer of amorphous silica and carbon which introduces a weak link near the fiber surface [1, 3-6, 24, 25]. The fibers should be pretreated to clean their surface and achieve a strong FI-bonding. Another example is the SiC + C fibers (Hi-Nicalon), whose microstructure is not fully stabilized after processing and whose surface may also contain some oxygen. During CVI-processing, fibers undergo a post-shrinkage weakening the FI-interface.   Again, the fibers should be pretreated. Finally, stoichiometric SiC fibers (Hi-Nicalon type S, HNS, or Tyranno SA, TSA), fabricated at higher temperatures are assumed to be dimensionally stable at composite processing temperatures. Further, their surface consists of free carbon (resulting from SiC decomposition) [26]. Hence, their bonding with PyC-interphase is expected to be relatively strong.

Finally, the roughness of fiber surface, which is low for Nicalon and Hi-Nicalon but significant for stoichiometric fibers, adds a mechanical contribution to the FI-bonding in a transition zone where the nanometric graphene layers stacks become progressively parallel to fiber axis [1, 7].

2.3- SiC (Nicalon)/PyC/SiC: a case history

Studies on SiC/PyC/SiC (CVI) fabricated with Nicalon or Hi-Nicalon fibers (as-received or pretreated), clearly show the positive effect of FM-interfacial design on material properties [1-4]. As shown in Fig. 1, tensile curves for the composites with a single 500 nm PyC-interphase (samples I and J) exhibit extended non-linear domains related to damaging phenomena, with high failure strains. However, composite with pre-treated fibers (sample J) is much stronger.

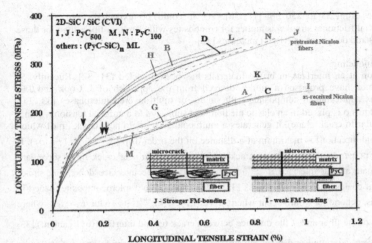

Fig. 1: Tensile curves at ambient of 2D-SiC/SiC (CVI) fabricated from Nicalon fibers with different interphases: single PyC-layers or (PyC-SiC)$_n$ ML-interphases (adapted from ref. [1]).

The shapes of the curves are different: continuously convex for the composite with pretreated fibers (sample J) and with a plateau-like feature for that with as-received fibers (sample I). Matrix microcracks are deflected near fiber surface (I-inset, right) for the latter, at the level of the weak carbon/silica interface and within the PyC-interphase (in a diffused manner and over short distance, J-inset, left) for the former. Here, the PyC-interphase does play its role of mechanical fuse. These two microcrack deflection modes correspond to very different FM-coupling, (weak in I and stronger in J). Interestingly, toughness of the composites fabricated with pretreated fibers as well as their fatigue resistance in tensile cyclic loading are higher when it is the PyC-interphase which is the active mechanical fuse (strong bonding) [1, 2, 27]. Similar results have been reported for composites fabricated from Hi-Nicalon fibers and PyC-interphase.

2.4- Influence of PyC-interphase thickness

PyC-layer thickness, e(PyC), has an influence on the mechanical properties of SiC/PyC/SiC composites, through thermal residual stresses (TRS) and fiber surface roughness.

In SiC/PyC/SiC (CVI) fabricated at $\approx$ 1000°C using Nicalon or Hi-Nicalon fibers, the coefficient of thermal expansion of the fibers is lower than that of the matrix, which results in compressive radial TRS at the FM-interface and reinforces FM-bonding. Further, the fiber surface is very smooth. Hence, increasing e(PyC) relaxes radial compressive TRS and lower interfacial shear stress. Experimental studies have shown that the mechanical properties go through an optimum for e(PyC) = few 100 nm [11, 15, 28]. P. Dupel et al. have reported that the tensile properties of 1D-SiC (NCG)/PyC/SiC (P-CVI) minicomposites were optima for e(PyC) = 220 nm. The calculated radial TRS in the interphase was compressive (enhancing thus the FM-bonding) at low e(PyC) values and tensile (favoring FM-debonding) for e(PyC) > 400 nm [29].

In SiC/SiC fabricated with stoichiometric fibers, the situation is different since the CTEs of the main constituents are now similar but the fiber surface is highly crystalline and rough. The mechanical properties of 2D-SiC/PyC/SiC (CVI) are either constant or slightly dependent on interphase thickness, when e(PyC) increases from 25 to 250 nm [17, 30]. As it could be expected from fiber roughness, the interfacial frictional stress τ_f is higher for composites with Tyranno fibers than for those with Hi-Nicalon fibers and decreases as e(PyC) increases [12].

2.5- Crack deflection modeling

Crack deflection at an interface in brittle materials has been modeled [31, 32]. Recently, S. Pompidou and J. Lamon have proposed a model, derived from the approach of J. Cook and J.E. Gordon [32], which is applicable to composites with single- or multi-layered interphases [33, 34]. When a crack of tip radius ρ is placed in an elastic medium and subjected to a uniaxial tension σ_{zz} (in a direction z perpendicular to crack plane), it generates a multiaxial stress field near crack tip of which σ_{rr} component (in radial direction) is maximum at a distance on the order of ρ ($\sigma_{rr}^{max} = \sigma_{rr}(r = \rho)$). If an interface is placed perpendicular to primary crack extension direction near crack tip, a secondary local crack may nucleate at that interface if $\sigma_{rr}^{max} > \sigma_i^c$, where σ_i^c is the interface debonding stress. Deflection results from coalescence of both cracks [32]. When applied to a microcomposite loaded in tension along fiber axis, debonding would occur when: $\sigma_i^c / \sigma_f^c \leq \sigma_{rr}^{max} / \sigma_{zz}^{max}$ (with r > l) , where σ_f^c is the failure stress of the fiber and, l, the distance between crack tip and interface (or ligament) [33, 34]. σ_{rr}^{max} and σ_{zz}^{max} were computed and their ratio plotted vs the Young's moduli ratio E_2 / E_1, as shown in Fig. 2. The domain under this master curve corresponds to the debonding situation and that above to conditions where debonding cannot occur. The curve exhibits a maximum corresponding to the highest debonding potential. Conversely, when E_2/E_1 decreases and tends to zero, debonding becomes quite impossible. But, the crack can be arrested. Failure of the reinforcing material depends on its strength versus stress operating: σ_f^c vs σ_{zz}^{max} .

The model has been applied to SiC/PyC/SiC to examine crack deflection probability at a given interface or within the interphase, to show the influence of fiber pretreatment and to discuss the effect of graphene layer orientation in the interphase. Deflection at first interface (SiC_m/PyC_i) is very unlikely since for the related E_2/E_1 ratio (≈ 0.07), the value of σ_i^c should be extremely low (Fig. 2) in accordance with experiments. By contrast, deflection at second interface (PyC_i/SiC_f) is most likely since for the corresponding E_2/E_1 ratio (10 for Hi-Nicalon) the width of the debonding domain (debonding potential) is very large. This is the most frequently observed case (weak FM-bonding). If the fiber has been pretreated to strengthen the FM-bonding, the representative point may move above the master curve, with debonding no longer occurring at that interface. However, it may take place within the PyC-interphase, i.e. at a PyC_i/PyC_i interface which shows for a E_2/E_1 value of 1, a still significant deflection potential (Fig. 2). This is the situation observed, for pretreated Nicalon fiber [1-4]. If now the graphene layers are deposited perpendicular to fiber surface, crack deflection within the PyC-interphase becomes no longer possible since the ratio $\sigma_{PyC//}^c / \sigma_{PyC\perp}^c \approx 2.17$ is well above the master curve for $E_2/E_1 = 1$.

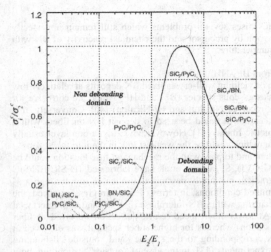

Fig.2: Values of σ_i^c / σ_2^c ratio provided by the master curve for various fiber/matrix and fiber/interphase/matrix systems (cracked material (material 1) cited second) (adapted from ref. [34]).

3- LAYERED INTERPHASES FOR SiC/SiC EXPOSED TO OXIDIZING ATMOSPHERE

Pyrocarbon is oxidation prone even at temperature as low as $\approx 500°C$, its oxidation resulting in the formation of gaseous oxides (active oxidation) and degradation of FM-coupling [35, 36]. Two approaches have been selected to solve this problem relying on self-healing (or self-sealing) mechanisms by condensed oxides (passive oxidation). The first one is based on single layer interphases containing boron whereas in the second, part of PyC is replaced, in so-called multilayered interphases, by SiC or TiC to reduce the thickness of each elementary PyC sublayer to a few 10 nm and to favor self-healing phenomena [9, 10].

3.1- Boron-doped pyrocarbon interphase

The addition of boron to pyrocarbon increases its graphitic character at low B-concentration and improves its oxidation resistance by blocking the so-called active sites and forming a fluid oxide (B_2O_3) in a temperature range (500-900°C) where the growth kinetics of silica is still too slow [37].

S. Jacques et al. have studied the influence of B-doped PyC-interphase on the oxidation resistance of 1D-SiC/SiC (CVI) microcomposites with pretreated Nicalon fibers [38]. Their interphases contain up to 30 at.% B. They showed, as expected, that the microtexture of the PyC-interphase was significantly improved at low B-addition, 8 at.% B, and degraded beyond this value (the interphase becoming amorphous). More importantly, lifetime in tensile static fatigue (beyond PL) in air at 700°C was dramatically improved as the B-content was raised, the best results being observed for graded interphase. Crack deflection and failure occur within the interphase at a location where the interphase microtexture and graphene layer orientation were optimal (at $\approx$ 8 at. % B).

3.2- Boron nitride interphases

The use of BN-interphase in SiC/SiC raises several problems which still remain imperfectly solved. They include the occurrence of corrosion by precursor and the chemical reactivity of BN with oxygen and moisture when prepared at low temperature.

3.2.1- BN-interphases as deposited by CVI from BF_3-NH_3

BF_3-NH_3 precursor has the advantage of yielding well-crystallized BN deposits at relatively low temperature [6, 39]. Unfortunately, it involves gaseous species (BF_3 and HF) which are corrosive for SiC-based fibers and alter their strength (as received Nicalon and Hi-Nicalon) [40]. Conversely, this precursor is compatible with carbon substrates and it could be used to deposit BN on fibers with a carbon layer surface (pretreated or stoichiometric fibers) [41]. However, an extra carbon layer usually remains between the fiber and the BN-coating which could be the weakest link in the interfacial zone.

One way to solve the corrosion problem and to play with the mechanical fuse location could be to deposit BN in a temperature gradient (TG-CVI). S. Jacques et al. have fabricated 1D-SiC/BN/SiC minicomposites with a radial crystallinity gradient by simply passing a Hi-Nicalon tow through a 3-temperature zones furnace [42]. Under optimized conditions, in terms of fiber progression speed, the FM-bonding was strong (crack deflection occurring within BN-interphase (Fig. 3)) and both interfacial shear stress and tensile stress were high. At lower fiber speed, crack deflection occurred at fiber surface (as a result of some surface crystallization) whereas for higher fiber speed it was observed at the SiC_m/BN interface, these two scenarios corresponding to the "inside" and "outside" debonding reported by G.N. Morscher et al., in related experiments [43]. In the case of "outside" debonding, both the interfacial shear stress and tensile failure stress were lower but the lifetime in tensile static fatigue at 700°C in dry or wet air was dramatically improved (crack deflection occurring far from fiber surface) [42].

3.2.2- BN-interphases as deposited by CVI from BCl_3-NH_3-H_2

BCl_3-NH_3-H_2 precursor is usually preferred since it is much less corrosive [44-47]. In principle, BN could be deposited at temperature as low as 700°C. However, under such mild conditions, it is nanoporous, poorly organized and highly reactive. Hence, the processing temperature should be increased [5, 44-47]. In the case of complex fiber architectures (nD-preforms), BN can be deposited at the highest temperature compatible with the ICVI-process ($\approx$ 1100°C) and further annealed at a temperature corresponding to the upper limit of the thermal stability domain of the fibers. An alternative is to deposit BN on fiber *tows*, which can be done at higher temperature (1400-1600°C), particularly for stoichiometric SiC-fibers [48]. As an example, BN deposited on a Tyranno SA tow at 1580°C was reported to be nearly stoichiometric, with an impurity content less than 5 at. %, highly crystallized and textured [49]. Another efficient way to improve the oxidation resistance of BN is to dope the precursor with a silane. The resulting BN(Si) deposit (15-40 wt. % Si), displayed an oxidation rate (at 1200-1500°C) 2-3 orders of magnitude lower than that for undoped BN [50]. Unfortunately, such Si-doped BN is amorphous and hence at variance with the first requirement of our layered interphase concept.

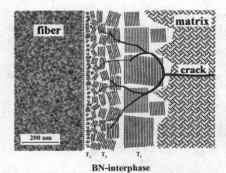

BN-interphase
Fig. 3: BN-interphase deposited from BCl_3-NH_3-H_2 on Hi-Nicalon fiber tow moving in a temperature gradient
($T_1 \leq 1100°C$; $T_2 = 1150°C$; $T_3 = 1250°C$) at medium residence time
(v = 2.5 m/h) (adapted from ref. [42]).

Generally speaking, the FM-bonding is relatively weak, owing to the occurrence of additional thin layers of silica or/and carbon at the fiber/BN-coating or/and at the SiC-matrix/BN-coating interfaces [44-47, 51, 52]. Matrix crack deflection occurs at these weak interfaces (mostly at the former) and not within the BN-interphase. Carbon is also present as single layer in composites fabricated with pre-treated fibers (Nicalon and Hi-Nicalon) or stoichiometric fibers. It can be chemically removed before BN-deposition. S. Le Gallet et al. have reported that such a treatment increased the interfacial shear stress, without achieving the high values required for crack deflection within the interphase [52]. Finally, crack deflection according to these different scenarios can be modeled, as already discussed for the composites with PyC-interphases (see Fig. 2) [33, 34].

3.2.3- BN-interphases as deposited by CVI from organometallics
Another way to reduce corrosion during BN-deposition is through the use of halogen-free organometallic precursors [53]. S. Jacques et al. have deposited BN-interphases from tris(dimethylamino) borane, B $[N(CH_3)_2]_3$ in H_2-NH_3 flow, on Hi-Nicalon fibers (NH_3 being used to avoid the occurrence of free carbon in the coating). The tensile curves of their minicomposites is characteristic of SiC/SiC with relatively strong FM-bonding, high matrix crack density at failure and high interfacial shear stress (here, $\tau = 230$ MPa). The latter is, to our knowledge, the highest value reported for SiC/BN/SiC composites. Finally, crack deflection did occur within the BN-interphase (Fig. 4), in accordance with the requirement of our layered interphase concept.

3.2.4- BN-interphases deposited on SiC fibers with in-situ generated BN-surface
The preceeding sections suggest that SiC-fibers with a carbon surface may not be the most appropriate for BN-deposition. An alternative would be to use SiC-fibers with a BN-surface. This can be achieved if a BN-film is formed at the expense of the fiber by radial diffusion coupled with reaction with an N-containing atmosphere [54-56].
A straightforward approach is to start with a stoichiometric SiC-fiber containing some boron (acting primarily as sintering aid) [26], as shown by M.D. Sacks and J.J. Brennan [54]. When such a fiber is treated at high temperature in an N-containing atmosphere, B-atoms diffuse radially from fiber core to react with nitrogen, yielding a strongly adherent BN-layer at fiber surface (typically, 100-200 nm thick). Further, diffusion in BN being anisotropic, the BN-coating grows with BN-layers

perpendicular to fiber surface and hence strongly bonded to the fiber. If now a BN-interphase is deposited on such a substrate by CVI, the BN-layers would have a tendency to be oriented, after some transition regime, parallel to the fiber surface [25, 27]. The so-called Sylramic-iBN and Super Sylramic-iBN fibers may have been developed on the basis of some related mechanism [55]. In the composites, crack deflection would occur either within the BN-interphase or at the BN-SiCm interface ("outside debonding"). These features could explain the good mechanical properties of these composites at high temperatures in oxidizing environment [43, 55, 57].

 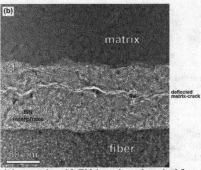

Fig. 4: 1D-SiC(HN)/BN/SiC(CVI) minicomposite with BN-interphase deposited from tris(dimethylamino)borane: matrix crack deflected within the BN-interphase, as seen by TEM (BF mode) at low (a) and high (b) magnifications (adapted from ref. [53]).

3.3- Multilayered (X-Y)$_n$ interphases

Multilayered (ML) interphases, (X-Y)$_n$, extend the concept of layered interphase to the nanometer scale, the interphase being now a stack of films of different materials X and Y, and the X-Y elementary sequence repeated n times. Their main advantage is that they can be highly tailored [1, 4, 5, 58]. As an example, the oxidation resistance of SiC/SiC could be improved by replacing PyC or BN interphases (100-200 nm thick) by (PyC-SiC)$_n$ or (BN-SiC)$_n$ ML-interphases in which the thickness of the oxidation-prone PyC or BN mechanical fuse is reduced to a few nanometers. This design criterion is based on oxygen gas phase diffusion consideration [35] and formation of healing condensed oxides (silica or/and boria) [13]. ML-interphases are deposited by CVI (switching from X to Y gaseous precursors) [58]. A key requirement is again a strong bonding between fiber surface and interphase.

3.3.1- (PyC-SiC)$_n$ multilayered interphases

Since the pioneering work of *C. Droillard et al.* [1, 2], ((PyC-SiC)$_n$ ML-interphases have been used in a variety of SiC/SiC [59-65]. Depending on CVI-conditions, SiC-sublayers are either microcrystallized (with rough SiC/PyC interfaces) or nanocrystallized with smooth SiC/PyC interfaces. Sublayer thickness is ranging from 3 to 100 nm for PyC and 10 to 500 nm for SiC while the number of PyC-SiC sequences is in the range of 3 to 10. The first material deposited on fiber surface is usually PyC but it could also be SiC in an attempt to strengthen the fiber/interphase bonding [64, 65]. Both MLs with constant sublayer thickness or graded sublayer thickness (on PyC or/and SiC) have been used [1, 65].

Replacing a PyC single interphase by a (PyC-SiC)$_n$ ML-interphase *does not change markedly* their tensile properties, as shown for composites fabricated with Nicalon fibers (Fig. 1) [1]. Tensile curves fall into two groups depending on whether the fibers are pretreated (strong FM-bonding) or not (weak FM-bonding). Similar conclusion can be drawn for SiC/SiC with Hi-Nicalon fibers [59, 60] or Tyranno SA stoichiometric fibers [65]. Crack deflection occurs at the fiber/interphase (or/and interphase/matrix) interface when the FM-bonding is weak (e.g. for as-received Nicalon or Hi-Nicalon fibers) and within the ML-interphase when it is strong enough (treated fibers). In this latter case, a matrix microcrack exhibits a dual propagation mode across the ML-interphase (Fig. 5) with an overall propagation path significantly lengthened [59, 60].

The lifetime of SiC/SiC with ML-interphase under load, at high temperature in air is improved with respect to their counterparts with single PyC-interphase [60, 66]. As an example, it increases from 2 to 48 hours for 2D-SiC (treated Nicalon)/SiC (CVI) composites, in tensile fatigue ($\sigma_{ap.}$ = 100 MPa) at 850°C, when a single PyC-interphase (50 nm thick) is replaced by a graded (PyC-SiC)$_4$ ML-interphase with e(PyC) = 50 nm and e(SiC) increasing from 50 to 150 nm when moving apart from fiber surface [66].

3.3.2- Other (X-Y)$_n$ multilayered interphases

At least two other ML-interphases, (BN-SiC)$_n$ and (PyC-TiC)$_n$, have been studied also with a view to improve the oxidation resistance. The potential advantage of the former lies in the fact that B-atoms are now present in the interphase, which favors crack-healing at intermediate temperatures. Unfortunately, experiments with pretreated Hi-Nicalon fibers have shown that the fiber/BN$_1$ bond is relatively weak, crack deflection occurring at that interface and not within the ML-interphase. Nevertheless, the lifetime in air at 700°C under load of minicomposites fabricated with Hi-Nicalon and a (BN$_{40}$-SiC$_{25}$)$_{10}$ ML interphase was significantly improved [9].

(PyC-TiC)$_n$ ML-interphases can also improve the oxidation resistance of SiC/SiC, although titanium oxides are not commonly regarded as healing oxides. Such interphases have been deposited on as-received Hi-Nicalon, according to a combination of conventional P-CVI (for PyC) and reactive P-CVI (for TiC) [67]. When the amount of TiC is low, each sublayer consists of a PyC-film reinforced with nanometric TiC-particles, which results in a strong FM-bonding. The lifetime of such minicomposites, under load in air at 700°C, is much higher (> 300 hours) than that (20 hours) for their counterparts with PyC single layer interphase. One of the reasons which could explain such unexpected behavior might be a strong interfacial bonding.

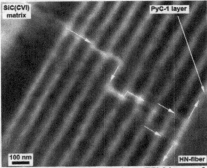

Fig. 5: 1D-SiC(HN)/SiC(CVI) with (PyC-SiC)$_{10}$ ML-interphase: TEM-image of a matrix microcrack deflected within the ML-interphase (adapted from ref. [60]).

3.4- Miscellaneous interphase materials

A few additional materials have been identified as potential interphase materials for SiC/SiC but with limited success up to now. This is the case for the ternary carbides (MAX-phases), such as Ti_3SiC_2 or Ti_3AlC_2, which display layered crystal structures. However, their deposition by CVD/CVI is difficult [68]. Further, they tend to grow with the layers perpendicular to the substrate surface and their ability to arrest/deflect a matrix crack in a CMC has not been formally established [69].

Oxides with layered crystal structures have been used as interphases in CMCs. This is the case for phyllosilicates (and related phyllosiloxides), such as mica fluorphlogopite $KMg_3 (AlSi_3)O_{10}F_2$ and the related phyllosiloxide $K(Mg_2Al)Si_4O_{12}$ [70]. However, their deposition (by sol gel process) is difficult, their thermal stability limited and their compatibility with SiC and SiC-CVI can be questioned. Other refractory oxides, such as zirconia, have also been considered but they do not have a layered crystal structure and hence, do not fall in our layered interphase concept [71].

Finally, porous interphases (also referred to as "pseudo-porous") consist of a mixture, at the nanometer scale, of a refractory material, such as SiC, with (fugitive) carbon [72]. However, such interphases do not display a marked anisotropic texture nor protect the fibers in an oxidizing environment and may undergo some sintering when exposed for a long time at high temperature.

4-   INTERPHASES IN SiC/SiC FOR HT-NUCLEAR REACTORS

SiC/SiC are potential structural materials for both fission and fusion high temperature nuclear reactors [15-17, 73]. This new and extremely demanding application raises specific constraints on the fibers, the matrix and the interphases.

4.1- SiC/SiC environment in HT nuclear reactors

The environment that would see SiC/SiC in e.g. gas cooled fast reactors is not so different from that they presently experience in advanced gas turbines, in terms of temperature, gas pressure and lifetime. However, they are not expected to see permanently oxidizing atmospheres. More importantly, they would be continuously exposed to intense irradiation by fast neutrons, α-particles and electromagnetic radiations. In high temperature advanced fission reactors, SiC/SiC will be exposed to moderately energetic neutrons (≈ 2 MeV) but at temperatures that could be higher than about 1200°C, whereas in Tokamak fusion reactor blankets, they will be irradiated by much more energetic neutrons (14.1 MeV) formed during the deuterium/titrium fusion reaction but likely at somewhat lower temperatures.

Such severe irradiation conditions are known to change (in a more or less dramatic manner) the structure of materials and hence their properties, as briefly discussed in the following sections for the SiC/SiC and their constituents, on the basis of literature data.

4.1.1- Irradiation of silicon carbide

Monolithic SiC undergoes a moderate swelling when irradiated by neutrons, as the result of amorphization or point defect formation at low temperature and cavity formation at high temperature. It first decreases as temperature is raised, passes through a minimum (0.2 to 0.4 vol% for a dose of 1-8 dpa) at 1100-1200°C, then increases to reach 1.5 vol% at 1600°C [17, 73]. SiC-matrix when deposited by CVI is pure, well crystallized and assumed to behave as monolithic SiC.

The effect of neutron irradiation on SiC-based fibers strongly depends on their composition and structure. On the one hand, stoichiometric fibers, which are well crystallized and with a small impurity content, also behave like monolithic SiC. On the other hand, Si-C-O (Nicalon-type) and SiC+C (Hi-Nicalon) fibers, prepared at lower temperature and poorly crystallized, undergo a permanent shrinkage [74].

As a result of this volume change mismatch upon neutron irradiation between SiC (CVI) matrix and 1st/2nd SiC fiber generations, debonding at FM-interface usually occurs with mechanical degradation [73, 75]. This key feature explains why stoichiometric SiC fibers are preferred for SiC/SiC to be used in nuclear reactors.

4.1.2- Neutron irradiation of interphase materials
Boron nitride, as an interphase, is poorly compatible with nuclear reactor environment [76, 77]. Firstly, $^{10}_5B$ isotope (present at a level of ≈ 20 at.% in natural boron) has an extremely high neutron capture cross section. Hence, the use of BN-interphase would suppose that it is deposited from $^{11}_5B$-enriched gaseous precursor. Secondly, boron nitride when neutron irradiated, undergoes nuclear reactions producing gaseous species (helium) and radioactive long life species (such as $^{14}_6C$). Further, since in a nuclear reactor the atmosphere is, in principle, not oxidizing, BN-interphases are, for all these reasons, not used in this field of application.

PyC-based interphases have been up to now the interphases of choice. However, they raise a problem of anisotropic dimensional change under neutron irradiation strongly depending on their degree of crystallization. Graphite and HOPG undergo a moderate shrinkage along the a-axis, i.e. parallel to graphene layer, and a swelling along the perpendicular c-axis. This anisotropy is strong at low termperature/high irradiation dose but it decreases as irradiation temperature is raised [78]. The behavior of turbostratic pyrocarbon is more complex. Although similar to that of graphite parallel to graphene layers, it first shrinks in a perpendicular direction at low irradiation dose and then swells [21, 79]. Since in a PyC-interphase graphene layers are preferably oriented parallel to the fiber surface, this dimensional change anisotropy may modify residual thermal stresses (particularly in radial direction) and alter the FM-bonding. Hence, the interphase may be again the vulnerable constituent of SiC/SiC when exposed in a prolonged manner to neutron irradiation. Potential solutions to overcome this difficulty are those already discussed in section 3, [16, 17, 80].

Finally, pseudo-porous SiC-interphases might also be an alternative as previously mentioned [16, 72, 80]. However, their dimensional stability under neutron irradiation is not well known. It may shrink (if it does contain enough free carbon), changing the residual stress field and favoring debonding.

4.2- Irradiated SiC/SiC
A compilation of flexural strength data reported by different authors, for a variety of SiC/PyC/SiC composites which have been neutron irradiated (in a broad temperature range: 200-1000°C) suggests that: (i) there is no loss in strength up to an irradiation dose of 10 dpa for materials fabricated with stoichiometric SiC-fibers, but conversely, (ii) the strength drops by $\approx 60\%$ almost linearly when irradiation dose increases up to 10 dpa for those with Si-C-O (Nicalon) or SiC + C (Hi-Nicalon) fibers [16, 73, 81]. This result is consistent with the dimensional change (permanent shrinkage) reported for these two latter fibers, which lowers the FM-bonding and load transfer. However, it would be more appropriate to rely on data determined using tensile tests. Hence, the analysis of the effects of neutron irradiation and interphase design on the mechanical properties of SiC/SiC should be pursued for composites with stoichiometric fibers, using tensile tests and analysis of microstructure-strength relations using appropriate models.

Fig. 6 shows tensile curve for SiC (HN-S)/PyC/SiC(CVI) composites recorded at ambient after neutron irradiation at 1000°C [19-21, 82]. For such a composite with a thick-PyC single layer interphase, the features of the tensile curve after irradiation show a weakening of the FM-bonding (with a plateau-like shape, broad hysteresis loops, and some residual strain after unloading). Further, the strain to failure is high (and close to that of dry tow under tension) as opposed to that of the unirradiated material (which seems to exhibit a premature failure) [82]. Irradiation at higher dose (up to ≈ 8 dpa at 800°C) of composites with still thicker PyC-interphase (≈ 700 nm) did not change markedly the tensile behavior after irradiation [20]. Since the thickness of the PyC-interphase is here extremely large (vs 100-200 nm in most SiC/SiC), the change in tensile behavior could be tentatively attributed to an evolution of the PyC-nanotexture during irradiation.

2D-SiC (HN-S)/PyC/SiC (CVI) composites, with much thinner PyC interphase (50-60 nm), display after neutron irradiation (750°C with dose up to 12 dpa), a very different behavior [19]. Their tensile curves (not shown in Fig. 6), before and after irradiation are very similar, with convex curvature up to failure, relatively narrow hysteresis loops and limited residual strain after unloading. These features suggest a relatively strong FM-bonding and little evolution of the interfacial zone during irradiation. However, in both cases, the strain-to-failure is low comparatively to that of the fiber. Hence, reducing the thickness of the PyC-interphase in composites fabricated with Hi-Nicalon S fibers, seems to enhance the stability of the FM-interphase bonding. It is noteworthy that such a conclusion cannot be drawn for those produced from Tyranno SA (where the FM-bonding seems to be higher and not to markedly depend on PyC-thickness) [17].

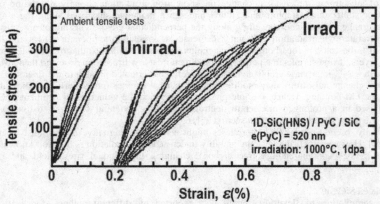

Fig. 6: Tensile curves of SiC/PyC/SiC composites with Hi-Nicalon type S SiC fibers and thick PyC interphase, before and after neutron irradiation, (adapted from ref. [82]).

Replacing the PyC single layer interphase by a (PyC-SiC)$_n$ ML-interphase (with (e(PyC) = 20 nm, e(SiC) = 100 nm and n = 5) yields relatively brittle composites, with extremely limited non-linear domain and very low strain at failure (≈ 0.1%). Their tensile curves after neutron irradiation (≈ 8 dpa ; 800°C) are similar to those of their un-irradiated counterparts [20]. Finally, composites with "pseudoporous" SiC interphases have also been irradiated and mechanically tested [16, 83].

5- CONCLUDING REMARKS

1- The interphases in SiC/SiC are ideally materials with a layered structure, in which the layers are parallel to the fiber surface, weakly bonded to one another but strongly adherent to the fiber. Matrix crack deflection occurs within the interphase in a diffuse manner and over short distances.

2- Anisotropic pyrocarbon is the interphase of   choice. It is deposited by CVI with graphene layers parallel to fiber surface. Achieving a strong bonding between the interphase and the fiber is not straightforward, and it may require a fiber pretreatment. The optimal PyC-thickness depends on both residual thermal stresses and fiber roughness. SiC/SiC with optimized PyC-interphase displays high load transfer and good mechanical properties under static or cyclic loadings in a broad temperature range. Unfortunately, pyrocarbon is oxidation-prone even at low temperatures.

3- BN layered interphase is more resistant to oxidation. However, its deposition by CVI with optimal structure and bonding to the fiber, is not straightforward. BF_3-NH_3 is a corrosive precursor whereas BCl_3-NH_3-H_2 requires high temperature to achieve high crystallinity and corrosion resistance. Corrosion problem could be solved through the use of TG-CVI or halogen-free organometallic precursor. Finally, achieving a strong fiber/BN-bonding remains a key issue. One way to solve this difficulty might be to use pretreated (stoichiometric) fibers with a BN-surface.

4- Another way to improve oxidation resistance of SiC/SiC is to reduce the thickness of PyC-interphase and to play with self-healing phenomena. SiC/SiC with $(X-Y)_n$ interphases (with X = PyC or BN and Y = SiC or TiC) displays improved lifetime under load in oxidizing atmospheres. The concept of multilayered material associated with self-healing phenomena, is still more efficient when extended to the matrix itself, oxygen being entrapped far from the fiber surface and interphase, with unmatched composite lifetimes.

5- The use of SiC/SiC in high temperature nuclear reactors is a relatively new field of application. BN being excluded from nuclear consideration, the best interphase material seems again to be pyrocarbon. However, pyrocarbon is not dimensionally stable under fast neutron irradiation, which may alter FM-bonding and composite mechanical properties. An approach to solve this problem is to lower the volume fraction of pyrocarbon in the interphase by either reducing its thickness (single or ML-interphases) to few nanometers or replacing pure carbon by C-SiC mixtures at the nanometer scale. Preliminary data suggest that the mechanical behavior of SiC/SiC with such interphases is not significantly degraded after neutron irradiation at 700-1000°C up to $\approx$ 10 dpa. However, more studies, including both representative mechanical testing and structure analysis of the interfacial zone (HR-TEM) should be pursued before considering that SiC/SiC fully meet all the requirements of this new and demanding application.

Acknowledgements

The authors acknowledge the contribution of all the researchers and engineers from both LCTS and its SPS and CEA partners, who have been involved in research program on interphases for SiC/SiC. They are indebted to J. Forget and C. Duhau for their assistance in the preparation of the present document.

REFERENCES

[1]C. Droillard, "Processing and characterization of SiC-matrix composites with C/SiC sequential interphase", *PhD-Thesis*, N° 913, Univ. Bordeaux 1, June 19, 1993.

[2]C. Droillard and J. Lamon, "Fracture toughness of 2D-woven SiC/SiC CVI-composites with multilayered interphases", *J. Am. Ceram. Soc.*, **79** [4] 849-858 (1996).

[3]R. Naslain, "The concept of layered interphases in SiC/SiC", *Ceram. Trans.*, **58** (1995) 23-39.

[4]R. Naslain, "The design of the fibre-matrix interfacial zone in ceramic matrix composites", *Composites Part A*, **29A** (1998) 1145-1155.

[5]R.A. Lowden, O.J. Schwartz and K.L. More, "Improved fiber coatings for Nicalon/SiC composites", *Ceram. Eng. Sci. Proc.*, **14** [7/8] 375-384 (1993).

[6]R. Naslain, O. Dugne, A. Guette, J. Sevely, C. Robin-Brosse, J.P. Rocher and J. Cotteret, "Boron nitride interphase in ceramic matrix composites", *J. Am. Ceram. Soc.*, **74** (1991) 2483-2488.

[7]R.J. Kerans, R.S. Hay, T.A. Parthasarathy and M.K. Cinibulk, "Interface design for oxidation-resistant ceramic composites", *J. Am. Ceram. Soc.*, **85** [11]- 2599-2632 (2002).

[8]S. Pasquier, J. Lamon and R. Naslain, "Tensile fatigue of 2D-SiC/SiC composites with multilayered (PyC-SiC)$_n$ interphases at high temperatures in oxidizing atmosphere", *Composites Part A*, **29A** (1998) 1157-1164.

[9]S. Bertrand, O. Boisron, R. Pailler, J. Lamon and R. Naslain, "(PyC-SiC)$_n$ and (BN-SiC)$_n$ nano-scale multilayered interphases by pressure pulsed-CVI", *Key. Eng. Mater.*, **164-165** (1999) 357-360.

[10]R. Naslain, R. Pailler, X. Bourrat, S. Bertrand and F. Lamouroux, "Non-oxide ceramic matrix composites with multilayered interphase and matrix for improved oxidation resistance", *Key. Eng. Mater.*, **206-213** (2002) 2189-2192.

[11]W. Yang, H. Araki, T. Noda, J.Y. Park, Y. Katoh, T. Hinoki, J. Yu and A. Kohyama, "Hi-Nicalon fiber-reinforced CVI-SiC matrix composites: I- Effects of PyC and PyC-SiC multilayers on the fracture behaviors and flexural properties", *Mater. Trans.*, **43** [10] 2568-2573 (2002).

[12]W. Yang, A. Kohyama, Y. Katoh, H. Araki, J. Yu and T. Noda, "Effect of carbon and silicon carbide/carbon interlayers on the mechanical behaviour of Tyranno SA-fiber reinforced silicon carbide-matrix composites", *J. Am. Ceram. Soc.*, **86** [5] 851-856 (2003).

[13] F. Lamouroux, S. Bertrand, R. Pailler, R. Naslain and M. Cataldi, « Oxidation resistant carbon fiber reinforced ceramic-matrix composites », *Composites Sci. Technol.*, **59** (1999) 1073-1085.

[14]F.A. Christin, "A global approach to fiber architectures and self-sealing matrices: from research to production", *Int. J. Appl. Ceram. Technol.*, **2** [2] 97-104 (2005).

[15]L.L. Snead, R.H. Jones, A. Kohyama and P. Fenici, "Status of silicon carbide composites for fusion", *J. Nucl. Mater.*, **233-237** (1996) 26-36.

[16]R.H. Jones, L. Giancarli, A. Hasegawa, Y. Katoh, A. Kohyama, B. Riccardi, L.L. Snead and W.J. Weber, "Promise and challenges of SiC/SiC composites for fusion energy applications", *J. Nucl. Mater.*, **307-311** (2002) 1057-1072.

[17]Y. Katoh, L.L. Snead, C.H. Henager Jr., A. Hasagawa, A. Kohyama, B. Riccardi and H. Hegeman, "Current status and critical issues for development of SiC composites for fusion applications", *J. Nucl. Mater.*, **367-370** (2007) 659-671.

[18]G. Newsome, L.L. Snead, T. Hinoki, Y. Katoh and D. Peters, "Evaluation of neutron irradiated silicon carbide and silicon carbide composites", *J. Nucl. Mater.*, **371** (2007) 76-89.

[19]K. Ozawa, T. Nozawa, Y. Katoh, T. Hinoki and A. Kohyama, "Mechanical properties of advanced SiC/SiC composites after neutron irradiation", *J. Nucl. Mater.*, **367-370** (2007) 713-718.

[20]Y. Katoh, T. Nozawa, L.L. Snead and T. Hinoki, "Effect of neutron irradiation on tensile properties of unidirectional silicon carbide composites", *J. Nucl. Mater.*, **367-370** (2007) 774-779.

[21]T. Nozawa, Y. Katoh and L.L. Snead, "The effects of neutron irradiation on shear properties of monolayered PyC and multilayered PyC/SiC interfaces of SiC/SiC composites", *J. Nucl. Mater.*, **367-370** (2007) 685-691.

[22]X. Bourrat, B. Trouvat, G. Limousin, G. Vignoles and F. Doux, « Pyrocarbon anisotropy as measured by electron diffraction and polarized light », *Carbon*, **15** [1] 92-101 (2000).

[23]P. Dupel, X. Bourrat and R. Pailler, « Structure of pyrocarbon infiltrated by pulse-CVI », *Carbon*, **33** [9] 1193-1204 (1995).

[24]M.H. Rawlins, T.P. Nolan, D.P. Stinton and R.A. Lowden, "Interfacial characterization of fiber-reinforced SiC-composites exhibiting brittle and toughened traction behavior", *Mater. Res. Soc. Symp. Proc.*, **78** (1987) 223-230.

[25]M. Monthioux and D. Cojean, "Microtextures of interfaces related to mechanical properties in ceramic fiber reinforced ceramic matrix composites", in *Proc. ECCM-5* (A.R. Bunsell et al., eds.), pp. 729-734, EACM, Bordeaux, 1992.

[26]S.M. Dong, G. Chollon, C. Labrugere, M. Lahaye, A. Guette, J.L. Bruneel, M. Couzi, R. Naslain and D.L. Jiang, "Characterization of nearly stoichiometric SiC ceramic fibers", *J. Mater. Sci.*, **36** [10] 2371-2382 (2001).

[27]J.M. Jouin, J. Cotteret and F. Christin, "SiC/SiC interphase: case history" in *Designing Ceramic Interfaces* II (S.D. Peteves, ed.), pp. 191-203, CEC, Luxembourg, 1993.

[28]E. Lara-Curzio, M.K. Ferber and R.A. Lowden, "The effect of fiber coating thickness on the interfacial properties of a continuous fiber ceramic matrix composite", *Ceram. Eng. Sci. Proc.*, **15** [5] 989-1000 (1994).

[29]P. Dupel, J-L. Bobet, R. Pailler and J. Lamon, "Effect of PyC-interphases deposited by pulsed-CVI on the mechanical properties of unidirectional composite materials" (in French), *J. Phys. III France*, **5** (1995) 937-961.

[30]T. Hinoki, E. Lara-Curzio and L.L. Snead, "Mechanical properties of high purity SiC fiber-reinforced CVI-SiC matrix composites", *Fusion Sci. Technol.*, **44** (2003) 211-218.

[31]M.Y. He and J.W. Hutchinson, "Crack deflection at an interface between dissimilar elastic materials", *Int. J. Solids Struct.*, **25** [9] 1053-1067 (1989).

[32]J. Cook and J.E. Gordon, "A mechanism for the control of crack propagation in all-brittle systems", *Proc. Roy. Soc.*, **28A** (1964) 508-520.

[33]J. Lamon and S. Pompidou, "Micromechanics-based evaluation of interfaces in ceramic matrix composites", *Adv. Sci. Technol.*, **50** (2006) 37-45.

[34]S. Pompidou and J. Lamon, "Analysis of crack deflection in ceramic matrix composites and multilayers based on the Cook and Gordon mechanism" *Composites Sci. Technol.*, **67** (2007) 2052-2060.

[35]L. Filipuzzi and R. Naslain, "Oxidation mechanisms and kinetics of 1D-SiC/C/SiC composites. II- Modeling", *J. Am. Ceram. Soc.*, **77** [2] 467-480 (1994).

[36]C. F. Windisch Jr., C. Henager Jr., G.D. Springer and R.H. Jones, "Oxidation of the carbon interface in Nicalon fiber reinforced silicon carbide composite", *J. Am. Ceram. Soc.*, **80** [3] 569-574 (1997).

[37]D.W. McKee, C.L. Spiro and E.J. Lamby, "The effects of boron additive on the oxidation behavior of carbons", *Carbon*, **22** [6] 507-511 (1984).

[38]S. Jacques, A. Guette, F. Langlais and R. Naslain, « C(B) materials as interphases in SiC/SiC model microcomposites », *J. Mater. Sci.*, **32** [4] 983-988 (1997).

[39] F. Rebillat, A. Guette, R. Naslain and C. Robin-Brosse, "Highly ordered BN obtained by LP-CVD", *J. Europ. Ceram. Soc.*, **17** (1997) 1403-1414.

[40]F. Rebillat, A. Guette and C. Robin-Brosse, "Chemical and mechanical alterations of SiC Nicalon fiber properties during the CVD/CVI process for boron nitride", *Acta Mater.*, **47** [5] 1685-1696 (1999).

[41]F. Rebillat, A. Guette, L. Espitalier and R. Naslain, « Chemical and mechanical degradation of Hi-Nicalon and Hi-Nicalon S fibers under CVD/CVI BN processing conditions », *Key Eng. Mater.*, **164-165** (1999) 31-34.

[42]S. Jacques, A. Lopez-Marure, C. Vincent, H. Vincent and J. Bouix, « SiC/SiC minicomposites with structure-graded BN interphases », *J. Europ. Ceram. Soc.*, **20** (2000) 1929-1938.

[43]G.N. Morscher, H.M. Yun, J.A. DiCarlo and L.T. Ogbuji, "Effect of a boron nitride interphase that debonds between the interphase and the matrix in SiC/SiC composites", *J. Am. Ceram. Soc.*, **87** [1] 104-112 (2004).

[44]K.L. More, K.S. Aisley, R.A. Lowden and H.T. Lin, "Evaluating the effect of oxygen content in BN interfacial coatings on the stability of SiC/BN/SiC composites", *Composites: Part A*, **30** (1999) 463-470.

[45]M. Leparoux, L. Vandenbulcke, S. Goujard, C. Robin-Brosse and J.M. Domergue, "Mechanical behavior of 2D-SiC/BN/SiC processed by ICVI", *Proc. ICCM-10*, Vol. IV, pp. 633-640, Woodhead Publ., Abington Cambridge, 1995.

[46]M. Leparoux, L. Vandenbulcke, V. Serin and J. Sevely, "The interphase and interface microstructure and chemistry of isothermal/isobaric chemical vapour infiltration SiC/BN/SiC composites: TEM and electron energy loss studies", *J. Mater. Sci.*, **32** (1997) 4595-4602.

[47]S. LeGallet, F. Rebillat, A. Guette and R. Naslain, "Stability in air at ambient temperature of BN-coatings processed from BCl$_3$-NH$_3$-H$_2$ gas mixtures", in *High Temperature Ceramic Matrix Composites* (W. Krenkel et al., eds.), pp. 187-198, Wiley-VCH, Weinheim, 2001.

[48]G.N. Morscher, D.R. Bryant and R.E. Tressler, "Environmental durability of BN-based interphases (for SiC$_f$/SiC$_m$ composites) in H$_2$O containing atmospheres at intermediate temperatures", *Ceram. Eng. Sci. Proc.*, **18** [3] 525-534 (1997).

[49]M. Suzuki, Y. Tanaka, Y. Inoue, N. Miyamoto, M. Sato and K. Goda, "Uniformization of boron nitride coating thickness by continuous chemical vapor deposition process for interphase of SiC/SiC composites", *J. Ceram. Soc. Japan*, **111** [12] 865-871 (2003).

[50]A.M. Moore, H. Sayir, S.C. Farmer and G.N. Morscher, "Improved interface coatings for SiC fibers in ceramic composites", *Ceram. Eng. Sci. Proc.*, **16** [4] 409-416 (1995).

[51]M. Leparoux, L. Vandenbulcke, V. Serin, J. Sevely, S. Goujard and C. Robin-Brosse, "Oxidizing environment influence on the mechanical properties and microstructure of 2D-SiC/BN/SiC composites processed by ICVI", *J. Europ. Ceram. Soc.*, **18** (1998) 715-723.

[52]S. LeGallet, F. Rebillat, A. Guette, X. Bourrat and F. Doux, "Influence of a multilayered matrix on the lifetime of SiC/BN/SiC minicomposites", *J. Mater. Sci.*, **39** (2004) 2089-2097.

[53]S. Jacques, B. Bonnetot, M-P. Berthet and H. Vincent, "BN-interphase processed by LP-CVD from tris(dimethylamino) borane and characterized using SiC/SiC minicomposites", *Ceram. Eng. Sci. Proc.*, **25** [4] 123-128 (2004).

[54]M.D. Sacks and J.J. Brennan, "Silicon carbide fibers with boron nitride coating", *Ceram. Eng. Sci. Proc.*, **21** [4] 275-281 (2000).

[55]H.M. Yun, D. Wheeler, Y. Chen and J. DiCarlo, "Thermomechanical properties of super Sylramic SiC fibers", in *Mechanical Properties and Performance of Engineering Ceramics and Composites* (E. Lara-Curzio, ed.), pp. 59-66, The Am. Ceram. Soc., Westerville, OH, 2005.

[56]L. Chen, H. Ye, Y. Gogotsi and M.J. McNallan, "Carbothermal synthesis of boron nitride coatings on silicon carbide", *J. Am. Ceram. Soc.*, **86** [11] 1830-1837 (2003).

[57]H.M. Yun, J.Z. Gyekenyesi, Y.L. Chen, D.R. Wheeler and J.A. DiCarlo, "Tensile behavior of SiC/SiC composites reinforced with treated Sylramic SiC fibers", *Ceram. Eng. Sci. Proc.*, **22** [3] 521-531 (2001).

[58]R. Naslain, R. Pailler, X. Bourrat, J-M. Goyheneche, A. Guette, F. Lamouroux, S. Bertrand and A. Fillion, « Engineering non-oxide CMCs at the nanometer scale by pressure-pulsed CVI », in *High Temperature Ceramic Matrix Composites 5* (M. Singh et al. eds.), pp. 55-62, The Am. Ceram. Soc., OH, 2004.

[59] S. Bertrand, P. Forio, R. Pailler and J. Lamon, "Hi- Nicalon/SiC minicomposites with (Pyrocarbon/SiC)$_n$ nanoscale multilayered interphases", *J. Am. Ceram. Soc.*, **82** [9] 2465-2473 (1999).

[60] S. Bertrand, R. Pailler and J. Lamon, "Influence of strong fiber/coating interfaces on the mechanical behavior and lifetime of Hi-Nicalon/(PyC/SiC)$_n$/SiC minicomposites", *J. Am. Ceram. Soc.*, **84** [4] 784-794 (2001).

[61] S. Bertrand, C. Droillard, R. Pailler, X. Bourrat and R. Naslain, "TEM structure of (PyC/SiC)n multilayered interphases in SiC/SiC composites", *J. Europ. Ceram. Soc.*, **20** (2000) 1-13.

[62] Y. Katoh, A. Kohyama, T. Hinoki, W. Yang and W. Zhang, "Mechanical properties of advanced SiC fiber-reinforced CVI-SiC composites", *Ceram. Eng. Sci. Proc.*, **21** [3] 399-406 (2000).

[63] T.M. Besmann, E.R. Kupp, E. Lara-Curzio and K.L. More, "Ceramic Composites with multilayer interface coatings", *J. Am. Ceram. Soc.*, **83** [12] 3014-3020 (2000).

[64] T. Hinoki, W. Yang, T. Nozawa, T. Shibayama, Y. Katoh and A. Kohyama, "Improvement of mechanical properties of SiC/SiC composites by various surface treatments of fibers", *J. Nucl. Mater.*, **289** (2001) 23-29.

[65] T. Taguchi, T. Nozawa, N. Igawa, Y. Katoh, S. Jitsukawa, A. Kohyama, T. Hinoki and L.L. Snead, "Fabrication of advanced SiC fiber/F-CVI SiC matrix composites with SiC/C multi-layer interphase", *J. Nucl. Mater.*, **329-333** (2004) 572-576.

[66] S. Pasquier, "Thermomechanical behaviour of SiC/SiC composite with a multilayered interphase: effect of environment", *PhD-Thesis, n°1727*, Univ. Bordeaux 1, Sept. 15, 1997.

[67] O. Rapaud, S. Jacques, H. Di-Muro, H. Vincent, M-P. Berthet and J. Bouix, « SiC/SiC minicomposites with (PyC/TiC)$_n$ interphases processed by presure-pulsed reactive CVI », *J. Mater. Sci.*, **39** (2004) 173-180.

[68] C. Racault, F. Langlais, R. Naslain and Y. Kihn, "On the chemical vapour deposition of Ti$_3$SiC$_2$ from TiCl$_4$-SiCl$_4$-CH$_4$-H$_2$ gas mixtures, Part II: An experimental approach", *J. Mater. Sci.*, **29** (1994) 3941-3948.

[69] S. Jacques and H. Fakih, "(SiC/Ti$_3$SiC$_2$)$_n$ multilayered coatings deposited by CVD", *Adv. Sci. Technol.*, **45** (2006) 1085-1090.

[70] R. Naslain, G. Demazeau and P. Reig, "Phyllosiloxides: new layered oxides as potential interphase materials", *Ceram. Trans.*, **79** (1996) 53-61.

[71] W.Y. Lee, E. Lara-Curzio and K.L. More, "Multilayered oxide interphase concept for ceramic-matrix composites", *J. Am. Ceram. Soc.*, **81** [3] 717-720 (1998).

[72] L.L. Snead, M.C. Osborne, R.A. Lowden, J. Strizak, R.J. Shinavski, K.L. More, W.S. Eatherly, J. Bailey and A.M. Williams, "Low dose irradiation performance of SiC interphase in SiC/SiC composites", *J. Nucl. Mater.*, **253** (1998) 20-30.

[73] A. Kohyama, "CMC for nuclear applications", in *Ceramic Matrix Composites* (W. Krenkel ed.), Chap. 15, pp. 353-384, Wiley-VCH, Weinheim, 2008.

[74] T. Hinoki, L.L. Snead, Y. Katoh, A. Hasegawa, T. Nozawa and A. Kohyama, "The effect of high dose/high temperature irradiation on high purity fibers and their silicon carbide composites", *J. Nucl. Mater.*, **307-311** (2002) 1157-1162.

[75] T. Hinoki, Y. Katoh and A. Kohyama, "Effect of fiber properties on neutron irradiated SiC/SiC composites", *Mater. Trans.*, **43** [4] 617-621 (2002).

[76] A. Hasegawa, A. Kohyama, R.H. Jones, L.L. Snead, B. Riccardi and P. Fenici, "Critical issue and current studies of SiC/SiC composites for fusion", *J. Nucl. Mater.*, **283-287** (2000) 128-137.

[77] T. Nozawa, T. Hinoki, Y. Katoh and A. Kohyama, "Effects of fibers and fabrication process on mechanical properties of neutron irradiated SiC/SiC composites", *J. Nucl. Mater.*, **307-311** (2002) 1173-1177.

[78] B.T. Kelly, W.H. Martin and P.T. Nettley, "Dimensional changes in pyrolytic graphite under fast-neutron irradiation", *Phil. Trans. Roy. Soc.*, **260** [A.1109] 37-49 (1996).

[79] J.L. Kaae, "The mechanical behaviour of Biso-coated fuel particles during irradiation. Part I: Analysis of stresses and strains generated in the coating of a Biso fuel particle during irradiation", *Nucl. Technol.*, **35** (1977) 359-367.

[80] B. Riccardi, L.G. Giancarli, A. Hasegawa, Y. Katoh, A. Kohyama, R.H. Jones and L.L. Snead, "Issues and advances in SiC/SiC composites development for fusion reactors", *J. Nucl. Mater.*, **329-333** (2004) 56-65.

[81] Y. Katoh, A. Kohyama, T. Hinoki and L.L. Snead, "Progress in SiC-based ceramic composites for fusion applications", *Fusion Sci. Technol.*, **44** (2003) 155-162.

[82] K. Ozawa, T. Hinoki, T. Nozawa, Y. Katoh, Y. Maki, S. Kondo, S. Ikeda and A. Kohyama, "Evaluation of fiber/matrix interfacial strength of neutron irradiated SiC/SiC composites using hysteresis loop analysis of tensile test", *Mater. Trans.*, **47** [1] 207-210 (2006).

[83] T. Hinoki, L.L. Snead, Y. Katoh, A. Kohyama and R. Shinavski, "The effect of neutron irradiation on the shear properties of SiC/SiC composites with varied interface", *J. Nucl. Mater.*, **283-287** (2000) 376-379.

RESEARCH AND DEVELOPMENTS ON C/C COMPOSITE FOR VERY HIGH TEMPERATURE REACTOR (VHTR) APPLICATION

Taiju Shibata, Junya Sumita, Taiyo Makita, Takashi Takagi, Eiji Kunimoto and Kazuhiro Sawa
High Temperature Fuel & Material Group, Japan Atomic Energy Agency,
4002 Oarai-machi, Ibaraki-ken, 311-1393 Japan

ABSTRACT

Japan Atomic Energy Agency (JAEA) carries out R&D on Very High Temperature Reactor (VHTR). Since core components in the VHTR will be used in severer conditions than in the High Temperature Engineering Test Reactor (HTTR), the application of heat-resistant ceramic composite material is important. Carbon fiber reinforced carbon matrix composite (C/C composite) and SiC fiber reinforced SiC matrix composite (SiC/SiC composite) are the major candidates for control rod substitute for metallic materials.

The scheme of the composite control rod development is categorized into the following phases; (1) Database establishment, (2) Design and (3) Demonstration test in the HTTR. In this paper, the development plans of (1) and (3) are described with some study results. For (1) Database establishment, it was shown that the irradiation effects on the C/C composite with graphitizing treatment exhibits similar trend with graphite. It would be possible to evaluate the irradiation effects based on the existing graphite irradiation database. Since JAEA has a set of graphite database established for the HTTR graphite components, it is expected that the graphite database can accelerate the composite control rod development. For (3) Demonstration test, the HTTR can irradiate large samples at high temperatures. It is possible to demonstrate the structural integrity of the mock-up of the composite control rod in the HTTR. The irradiation demonstration is necessary for the final phase of the development. The irradiation test conditions for the HTTR demonstration are explained.

INTRODUCTION

Very High Temperature Reactor (VHTR) of Generation-IV reactor is an advanced High Temperature Gas-cooled Reactor (HTGR) which is graphite-moderated and helium gas-cooled. It can provide high temperature helium gas of about 950 °C to the reactor outlet. It is possible to use this high temperature as the heat source not only for electrical power generation but also for hydrogen production. Japan Atomic Energy Agency (JAEA) developed the High Temperature Engineering Test Reactor (HTTR) which is a test reactor with thermal power at 30 MW and the first HTGR in Japan[1]. The first criticality of the HTTR was achieved in 1998, and the full power operation of 30 MW was attained in 2001 with a reactor outlet temperature of 850℃. The high-temperature operation of 950℃ was successfully done in 2004. A long-term high temperature operation test is planned in 2009. The nuclear heat utilization thermochemical hydrogen production by IS (Iodine-Sulfur) process will be demonstrated by coupling with HTTR[2,3]. JAEA carries out development of the VHTR based on the HTTR technologies.

The inherent and passive safety feature of the VHTR is also attractive. It is expected that the VHTR shows slow temperature change behavior in accident conditions because graphite components with a large heat capacity are used in the core. The HTTR has been demonstrating the excellent safety features through the series of safety demonstration tests since 2002. Reactivity insertion tests (control rod withdrawal tests) and gas circulators trip tests (by running down one and two out of three gas circulators) were successfully completed. A loss of forced cooling test (by running down all gas circulators), a vessel cooling system stop test and an off-normal load condition test of heat utilization system are under planning.

Since core components in the VHTR will be used in severer conditions than in the HTTR, it is important to develop heat-resistant ceramic composite material substitute for metallic materials. For example, ferritic superalloy Alloy 800H is used for the control rod sheath in the HTTR[4]. Its maximum allowable temperature to be used repeatedly after scrams is 900 °C. The heat-resistant ceramic composite, carbon fiber reinforced carbon matrix composite (C/C composite) and SiC fiber reinforced SiC matrix composite (SiC/SiC composite) are the major candidates substitute for the metallic materials to use at higher temperature. Their application for the control rod sheath is one of the key subjects for the VHTR development.

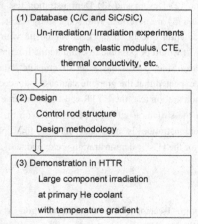

Fig. 1 Scheme of C/C and SiC/SiC composite control rod development.

JAEA studies on the application of the composite materials for the in-core components of VHTR [5-8]. The scheme of the control rod sheath development is categorized into the following phases shown in Fig. 1; (1) Database establishment, (2) Design and (3) Demonstration test in the HTTR. This paper first describes the outline of the VHTR development in Japan and then describes the plan of the composite control rod development. C/C composite is focused in the studies (1) and (3) in this study. Irradiation test results on the C/C composite are also discussed in this paper.

R&D ITEMS FOR VHTR DEVELOPMENT IN JAPAN

The details of some R&D items being necessary for the VHTR development are described in the followings.

Graphite Components

Figure 2 shows the cutaway drawing of the HTTR core[1]. The HTTR uses pin-in block type fuel element which is a hexagonal graphite block, 360 mm across flats and 580 mm in height. The fine-grained isotropic graphite IG-110 (Toyo Tanso Co.) is used for the hexagonal fuel, control rod guide, and replaceable reflector blocks. IG-110 is also used for the core support post. PGX graphite and ASR-0RB carbon are used for the core support components. JAEA established a design code for the HTTR core components which includes irradiation effect on IG-110 graphite material properties. The graphite and carbon material property database was established through experiments.

It is expected that these design code and database are available for the VHTR design. IG-110 graphite is also a major candidate for the core components in the VHTR. A special committee at Atomic Energy Society of Japan (AESJ) discussed the technical criteria for VHTR graphite components based on the HTTR design code. The discussion included graphite database, design, materials, and in-service inspection requirements. Extrapolation of irradiation data on graphite and application of fracture mechanics on the design code were also discussed. The discussion at the special committee was completed in March 2009. JAEA also joins the discussion on the establishment of the American Society of Mechanical Engineers (ASME) code for graphite core components.

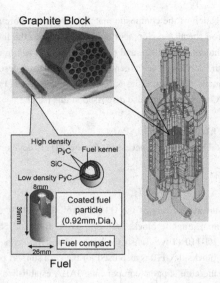

Fig. 2 Cutaway drawing of HTTR core[1].

Triso-coated Fuel Particle

HTTR uses Triso-coated fuel particles as shown in Fig. 2. The Triso-coated fuel particle consists of a micro spherical kernel of oxide or oxycarbide fuel and coating layers of porous pyrolytic carbon (buffer), inner dense pyrolytic carbon (IPyC), silicon carbide (SiC) and outer dense pyrolytic carbon (OPyC). The principal function of these coating layers is to retain fission products within the particle. Through the HTTR operation, fission gas release rate was very low, which means that the fuel has very high quality, i.e., very low as-fabricated failure fraction in a commercial scale.

Although SiC has excellent material properties, it gradually loses strength due to neutron irradiation and mechanical integrity at very high temperatures, especially above 1700 °C, by thermal dissociation. Zirconium carbide (ZrC) is known as a refractory and chemically stable compound. The ZrC coating layer is an advanced candidate to replace the SiC coating layer of the Triso-coated fuel particles for VHTR. The resulting particle is termed a ZrC Triso-coated fuel particle. JAEA has been studying the ZrC-coated fuel particle as an advanced particle. Coating conditions to obtain ZrC with uniform structure and with good stoichiometry, that is, C/Zr ratio of 1.0, were acquired by a large-scale with 0.2kg-batch size. Coating conditions of the ZrC layer have been optimized through temperature control in the coating process[9,10].

Control Rod Development

Ferritic superalloy Alloy 800H is used for the material of the metal parts of the control rod in the HTTR[1,4] as shown in Fig. 3. The maximum allowable temperature for the control rod to be used repeatedly after scrams is 900 ℃. To keep this limit, HTTR adopts the two-step control rods insertion method at a reactor scram. The control rods at the outer region of the core are inserted first and the control rods at the inner region are inserted later so as to keep the control rod temperature below 900 ℃ [4]. The heat-resistant ceramic composite, C/C and SiC/SiC composites are major candidates substitute for the metallic materials of control rod in order to avoid this complicated procedure. For this purpose, their application for the control rod sheath is one of the important subjects for the VHTR development.

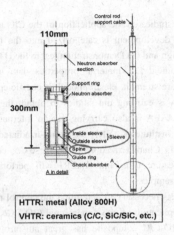

Fig. 3 Control rod structure of HTTR[1,4].

For the composite control rod development, it is important to obtain database including irradiation effects. It is possible to carry out irradiation test with small size specimens. However, it is difficult to demonstrate the integrity of the control rod structure by the mock-up in the HTGR condition, though it is quite important for the final stage of the control rod development. The HTTR has a capability to carry out various irradiation tests at very wide irradiation regions at elevated temperatures in the core. The real scale control rod can be irradiated in the HTTR core at high temperature. It means that it is possible to demonstrate control rod components with C/C and SiC/SiC composite in the HTTR condition.

Hydrogen Production Technology

A thermochemical hydrogen production cycle IS process has been developed step by step[11,12]. The IS process was verified with a lab-scale apparatus at a hydrogen production rate of 1 L/h in 1997. Further tests have been conducted. Continuous hydrogen production was successfully achieved by using a bench-scale test apparatus with a hydrogen production rate of about 31 L/h for 1 week in 2004. The demonstration of the IS process by coupling with HTTR is the goal of the HTTR project. The target is the hydrogen production rate is up to 1000 m^3/h. For the demonstration of the coupling, system integration technologies, i.e. safety evaluation and component tests are necessary.

R&Ds ON CONTROL ROD SHEATH WITH COMPOSITE

R&D Scheme

As shown in Fig. 1, JAEA studies on the application of the C/C and SiC/SiC composites for the VHTR control rod sheath. The development is categorized into the following phases. They are (1) Database establishment, (2) Design and (3) Demonstration test in the HTTR.

For the database establishment (1), material properties data in un-irradiated and irradiated conditions are necessary. They are strength, elastic modulus, coefficient of thermal expansion (CTE), thermal conductivity, etc. JAEA is carrying out studies to obtain these data for a several kinds of candidate composite materials. JAEA is also carrying out an international collaboration study with Korea Atomic Energy Research Institute (KAERI) for the un-irradiated material properties evaluation for some C/C composite materials. Characterizations of Japanese C/C composites are being conducted. JAEA will perform fracture feature tests and KAERI will perform oxidation examination and characterization on the oxidized samples.

For the design phase (2), it is necessary to determine the control rod structure. One of the major study items is focused on the application of two-dimensional (2D-) C/C composite for the control rod sheath application. Since the 2D-C/C composite has great anisotropy in thermal and mechanical properties in parallel and vertical to lamina directions, it is important to consider the anisotropy for component design. Connecting method for the component parts is one of the key technologies. Appropriate design methodology, e.g. fracture theory, for the fiber reinforced ceramics is not fully developed yet. It is also important issues for the C/C and SiC/SiC composite application.

For the demonstration test in the HTTR (3), it is the final stage for the composite control rod development. It is difficult to find the irradiation field which can irradiate real scale control rod structure at the HTGR condition except the HTTR. Although the irradiation flux of the HTTR is not high, it is possible to irradiate at primary helium-gas coolant condition with temperature gradient. It would be expected to carry out this demonstration test through international collaboration.

The details of above irradiation test results in (1) and the HTTR demonstration test in (3) are described in the followings.

Irradiation Test

(1) Experimental

PAN-based CX-270G grade (Toyo Tanso Co.) is one of the candidate materials for of 2D- C/C composite. It is graphitized at 2800 °C in its manufacturing process. Table 1 shows its typical material properties in comparison with IG-110 graphite. The samples were irradiated at 03M-47AS capsule in Japan Material Testing Reactor (JMTR) of JAEA to the neutron fluence of 8.2×10^{24} n/m^2 (E>1.0MeV) corresponding to 1.2 dpa (displacements per atom) at 600 °C. The irradiation-induced dimensional change and thermal diffusivity were measured[5]. Test methods are summarized in Table 2.

Table 1 Typical material properties of CX-270G and IG-110.

Grade	Bulk density (Mg/m)	Bending strength (MPa)	Tensile strength (MPa)	Compressive strength (MPa)	Elastic modulus (GPa)	Thermal expansion coefficient (10^{-6}/K)
CX-270G-pl	1.63		167	69	81	0.2
CX-270G-pp	1.63	133		89		10.8
IG-110	1.77	39	25	78	10	4.6

·pl: parallel to lamina direction, ·pp: perpendicular to lamina direction

Table 2 Test methods of CX-270G irradiation effects.

property	specimen size	measurement method
dimensional change	φ5x20 mm	micro meter
thermal diffusivity	φ10x2 mm	laser flash method from room temperature to 1400 °C, for every 100 °C

(2) Results

1) Dimensional change

Irradiation-induced dimensional change[5] is shown in Fig. 4. The dimensional change of the CX-270G at 600 °C irradiation is expressed by the following equations.

$$\Delta L / L = -aF \tag{1}$$

$a = 4.7 \times 10^{-2}$ (%/dpa), for perpendicular to lamina direction

$a = 4.4 \times 10^{-1}$ (%/dpa), for parallel to lamina direction

$\Delta L / L$ is change of dimension (%), a is a constant and F is fast neutron fluence(dpa).

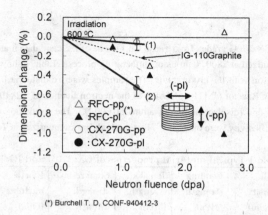

Fig.4 Irradiation-induced dimensional change of CX-270G.
(Evaluation lines (1) and (2) are respectively for perpendicular and parallel directions.)

It is obvious that the shrinkage for the parallel to lamina direction is much greater than that for the perpendicular to lamina direction. It was reported that irradiation causes shrinkage of fibers along their length and swelling in their diameter. Reported data for the PAN-based 2D- random fiber composite (RFC)[13] are also plotted in Fig. 4. Although the trend of our data is similar to that of the RFC, there is little difference in the amount of the change at 1.2 dpa. It is thought that the difference is derived from their structural geometries in lamina. The lamina of RFC is random fiber structure and CX-270G is plain-woven 2D-C/C composite. The design data of IG-110 graphite for the HTTR[1] is also drawn in the figure. The linearly decreasing trend of IG-110 is similar to CX-270G. It is due to the graphitization treatment for the CX-270G. It suggested that it would be possible to evaluate the irradiation effects of graphitized C/C composite based on the effects on the graphite. Existing graphite irradiation database would be used to evaluate the C/C composite.

2) Thermal conductivity

Thermal conductivity was calculated by the measured thermal diffusivity[5]. Figure 5 shows the thermal conductivity ratio of irradiated to un-irradiated samples of 2D-C/C composite CX-270G. The data for IG-110 are also plotted in the figure. We can see that the measurement temperature dependency of the conductivity is almost the same for the both directions of CX-270G and IG-110. It is thought that the irradiation effect on the graphitized CX-270G sample is almost the same with the graphite. It is reported that the heat transport in graphite takes place by lattice wave conduction (phonon conduction). Thermal conductivity of graphite reduces when it was irradiated, because phonons are scattered by irradiation defects in the graphite crystal structure. It is also thought that graphite irradiation database would be used to evaluate the graphitized C/C composite thermal properties. The ratio increases with the temperature and rapidly increases above 1200 °C for every sample. The difference of the increasing ratios is obvious in the figure. It would be due to the annealing recovery effect of the irradiation damage.

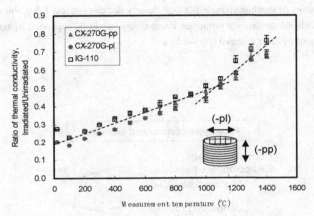

Fig.5 Irradiation-induced thermal conducutivity change of CX-270G.

JAEA has been carrying out the post irradiation experiments on the irradiated CX-270G samples. Coefficient of thermal expansion and Young's modulus by ultrasonic wave method were obtained. Additional data on the thermal conductivity was also obtained. The establishment of database and evaluation based on the graphite database is underway.

C/C Control Rod Demonstration Test in HTTR

The concept of the composite control rod structure is shown in Fig. 6. Major parts are sheath (inner/ outer sleeves) and connecting rod (spine). For the tubular parts, application of the 2D-C/C composite is candidate in the viewpoint of fabrication. For the connecting rod, oxidation resistant SiC/SiC composite is candidate material. Since the control rod has three-dimensional structures, stress concentration would be occurred in some points due to temperature gradient and irradiation-induced dimensional change in the reactor core.   It is necessary to study and demonstrate its integrity for the in-core use.

The HTTR can be used as the irradiation field and it is possible to demonstrate the integrity[14,15]. Figure 7 shows the irradiation capability of the HTTR[1,14]. A large-scale component irradiation test can be conducted on a high temperature condition. The temperature gradient is about 200 °C in the depth span of 2500 mm in an irradiation region[15]. It is possible to check the stress concentration by temperature gradient. Figure 8 shows estimated irradiation effects as a function of neutron irradiation fluence. Although the expected neutron fluence by the HTTR is not so high, the thermal conductivity would decrease to about 30% in the HTTR irradiation condition. It is because the thermal conductivity of graphitized materials significantly decreases at less 1 dpa. For the evaluation of the irradiation-induced mechanical properties change, they would be supported by the irradiation test with test specimens by other material test reactors.

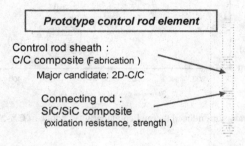

Fig.6 Concept of the composite control rod structure.

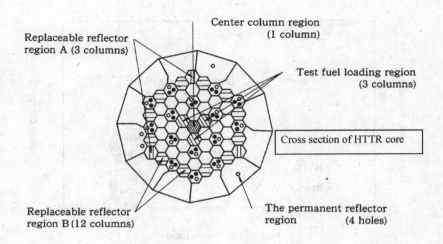

Fig.7 Irradiation performance of the HTTR[1,14].

Irradiation region	Max. irradiation volume (mm)	Max. neutron flux (n/m²/s)		Region temperature (℃)
		Fast (>0.18MeV)	Thermal (<2.38eV)	
Center column region	1 block size (*)	2×10^{17}	7×10^{17}	400-1100
Test fuel loading region	1 block size (*)	2×10^{17}	5×10^{17}	400-1100
Replaceable reflector region A	300(Dia.)x500(L)	2×10^{16}	4×10^{17}	400-800
Replaceable reflector region B	130(Dia.)x500(L)	2×10^{16}	4×10^{17}	400-800
Permanent reflector region	100(Dia.)x3000(L)	8×10^{14}	3×10^{17}	400-600

(*) Hexagonal block with 360(across flat)x580(H)mm
(**) Length at specimen

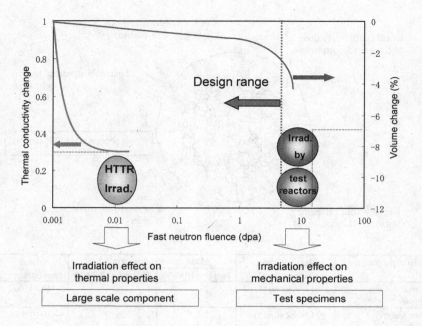

Fig.8 Estimated test conditions in the HTTR demonstration.

CONCLUDING REMARKS

The outline of VHTR development and the scheme of the composite control rod development were described. The scheme of the development is categorized into; (1) Database establishment, (2) Design and (3) Demonstration test in the HTTR. The followings were expressed with experimental data.

The irradiation effects of the graphitized C/C composite, dimensional change and thermal conductivity change, showed similar trend with that of graphite. It is effective to use existing graphite irradiation database for the graphitized C/C composite evaluation. Since JAEA has a set of graphite database for the HTTR graphite components, it is expected that the graphite database can accelerate the composite control rod development.

The HTTR can irradiate large samples at high temperatures. It is possible to demonstrate the structural integrity of the mock-up of the composite control rod in the HTTR being necessary for the final phase of the development. The irradiation effects on the thermal properties of the composite mock-up can be tested in the HTTR.

REFERENCES

[1] S. Saito, et al., Design of High Temperature Engineering Test Reactor (HTTR), JAERI 1332 (1994).

[2] K. Matsui, S. Shiozawa, M. Ogawa and X. L. Yan, Present Status of HTGR Development in Japan, Proc. 16th Pacific Basin Nuclear Conference (16PBNC), Aomori, Japan, Oct. 13-18, 2008, PaperID P16P1355

[3] S. Shiozawa, et al., Status of the Japanese Nuclear Hydrogen Program, ANS 2007 Annual Meeting, Embedded Topical on Safety and Technology of Nuclear Hydrogen Production, Control and Management" (ST-NH2), June24-28, 2007, Boston, MA, USA.

[4] Y. Tachibana, S. Shiozawa, J. Fukakur, F. Matsumoto and T. Araki, Integrity assessment of the high temperature engineering test reactor (HTTR) control rod at very high temperatures, Nucl. Eng. and Design 172, 93-102(1997).

[5] T. Shibata, J. Sumita, E. Kunimoto and K. Sawa, Characterization of 2D-C/C composite for application to in-core structure of Very High Temperature Reactor (VHTR), Proc. Carbon 2008, P0370, 13-18th, July, 2008, Nagano, Japan.

[6] T. Shibata, J. Sumita, S. Baba, M. Yamaji, et al., Tensile strength of two-dimensional C/C composite with its microstructure for nuclear application, Proc. M&P2005, Seattle, USA, CD-ROM, CMC-11(2005).

[7] S. Hanawa, J. Sumita, T. Shibata, M. Ishihara, et al., Stress analysis of two-dimensional C/C composite components for HTGR's core restraint mechanism, Proc. SMiRT 18, Beijing, China, CD-ROM, SMiRT18-C06-1(2005).

[8] T. Sogabe, M. Ishihara, et al., Development of the Carbon Fiber Reinforced Carbon-carbon Composite for High Temperature Gas-cooled Reactors, JAERI-Research 2002-026(2002).

[9] K. Sawa, Advanced Coated Particle Fuel - ZrC coated particle development in JAEA, EUROCOURSE on coated particle fuel, Petten, December 4-7, 2007.

[10] S. Uata, J. Aihara, A. Yasuda, H. Ishibashi, et al., Fabrication of uniform ZrC coating layer for the coated fuel particle of the very high temperature reactor, J. Nucl. Mater. 376, 146-151(2008).

[11] S. Kubo, et al., A Demonstration Study on a Closed-Cycle Hydrogen Production by Thermochemical Water-Splitting Iodine-Sulfur Process, Nucl. Eng. Des., 233, .347-354 (2004).

[12] J. Iwatsuki, et al., Design study of pilot test plant for hydrogen production by thermochemical water splitting IS process, 15th International Conference on Nuclear Engineering (ICONE-15), April 22-26, 2007, Nagoya, Japan, ICONE15-10178.

[13] Burchell T. D, CONF-940412-3.

[14] T. Shibata, T. Kikuchi, S. Miyamoto and H. Obata, Development of Irradiation Rig in HTTR and Dosimetry Method - I-I Type Irradiation Equipment -, JAERI-Tech 2002-097 (2002).

[15] T. Shibata, T. Kikuchi, S. Miyamoto and K. Ogura, Assessment of irradiation temperature stability of the first irradiation test rig in the HTTR, Nucl. Eng. and Design, 223, 133-143 (2003).

X-RAY TOMOGRAPHIC CHARACTERIZATION OF THE MACROSCOPIC POROSITY OF CVI SIC/SIC COMPOSITES—EFFECTS ON THE ELASTIC BEHAVIOR

L. Gélébart (1), C. Chateau (1-2), M. Bornert (2), J. Crépin (3), E. Boller (4)

(1) CEA Saclay, SRMA, France
(2) LMS-Ecole Polytechnique, France
(3) CDM-ENSMP, France
(4) ESRF Grenoble, ID19, France

ABSTRACT

In the context of the development of the next generation of nuclear reactors, SiC/SiC composites are candidate for structural applications. Because of their complex thermo-mechanical behaviour, due to their complex microstructure, a multi-scale approach is under development. An important microstructural parameter of the CVI composite is the complex distribution of the residual porosity inherent to the CVI process.

This paper focuses on the characterization of the macroporosity (the porosity among the tows) and on its effect on the thermo-mechanical behaviour. The experimental characterization of the macroporosity is performed using an X-ray tomography technique on the beamline ID19 at the ESRF synchrotron (France) with a resolution of $5.02^3 \mu m^3$.

The numerical 3D images are used to describe the distribution of macroporosity with respect to the position of the plies. It is clearly established that the stacking of the plies has a significant effect on the porosity distribution. As a consequence for the micromechanical modelling, a unique element that contains only one ply is not representative of the porosity distribution and is not sufficient to evaluate the "effective" mechanical properties: several volume elements, called "statistical volume elements", with at least 2 plies per volume element have to be used in order to account for the variability of the stacking of the plies.

Finally, such "statistical volume elements" (SVE) are directly extracted from the tomographic image and the "effective" elastic behaviour is evaluated from the average of the "apparent" behaviour evaluated on each SVE. In spite of their quite important size ($3.2 \times 3.2 \times 0.45 mm^3$), the "apparent" behaviours evaluated for each SVE exhibit important fluctuations.

1 – INTRODUCTION

SiC/SiC composites are potential candidate for structural applications in the next generation of nuclear reactors. To study their complex thermo-mechanical behaviour, due to their complex microstructure, a multi-scale approach is under development at CEA/SRMA. An important microstructural parameter of the CVI (Chemical Vapor Infiltration) composite is the complex distribution of the porosity which can be described at two different scales: the micro-porosity is observed within the tows among the fibres, the macro-porosity is observed among the tows. The effect of the micro-porosity has already been evaluated on the elastic behaviour of the tow: its complex geometry that results from the CVI process induces a highly anisotropic behaviour of the tow and a high level of stress concentrations[1].

This paper focuses on the characterization of the macro-porosity (*i.e.* among the tows) and on its effect on the anisotropy of the composite's elastic behaviour. The experimental characterization of the macro-porosity is performed using an X-ray tomography technique on the beamline ID19 at the ESRF synchrotron (in Grenoble, France). These results are used to study the distribution of the macro-porosity within the composite. The influence of the stacking of the plies on this distribution is emphasized and the consequences on the modelling are discussed. The numerical 3D images are then used to obtain an estimate of the "effective" elastic behaviour of the composite. This estimate is based on Finite Element calculations performed on several volume elements (called "statistical volume elements") extracted from the 3D image. The fluctuations observed on the "apparent" elastic behaviour (i.e. the behaviour evaluated for each "statistical volume element") are also discussed

2 – MATERIAL AND EXPERIMENTAL PROCEDURE

2.1 – Material

The material under investigation is a 2D SiC/SiC composite provided by SNECMA SPS. The material is elaborated from a fibrous preform on the base of yarns constituted by 500 Hi-Nicalon type S fibres. The preform consists of a stacking of 13 plies and each ply has a plain weave architecture. A 3D representation of the plain weave architecture is displayed on Figure 11a. A CVI is then used to deposit a 100nm interphase of pyrocarbon and then to densify the composite with SiC matrix. Because of the CVI process, the material can not be fully densified and a residual porosity appears within the composite. This porosity can be described at two different scales: the micro- and the macro-porosity. The micro-porosity is the porosity that takes place within the yarns, among the fibres. This porosity consists of closed elongated pores aligned in the direction of the yarn. A statistical analysis performed on different yarn cross-sections (Figure 1) reveals that the mean volume fraction of fibres is 55%, the mean volume fraction of micro-porosity is 4% and a typical size of the pores is 5μm (but the distribution is highly heterogeneous). The macro-porosity is the porosity that can be observed among the yarns. The 2D observations such as Scanning Electron Microscope (SEM) observations (Figure 2) are insufficient to describe the complexity of the macro-porosity which is generally an open porosity with a 3D complex shape. The only way to characterize this complex macro-porosity is to use 3D characterization techniques such as X-ray tomography.

Figure 1: distribution of the micro-porosity: cross-section of a yarn (SEM observation)

Figure 2: distribution of the macro-porosity: section of the composite (SEM observation)

2.2 – X-ray tomography

X-ray tomography is a non destructive technique which allows imaging a sample in 3 dimensions with micronic (and even submicronic) resolutions. This technique can be divided into 2 successive steps. The first one consists in recording a set of radiographs for different angular positions of the sample with respect to the X-ray beam. The second step uses an appropriate algorithm (filtered back projection) to reconstruct, from this set of radiographs, the spatial distribution of the absorption coefficient within the sample (Figure 3). As a consequence, the contrast observed in the 3D images is given by the spatial heterogeneity of the local absorption coefficient which can be correlated to the local density. For CVI SiC/SiC composites, the porosity induces highly contrasted images.

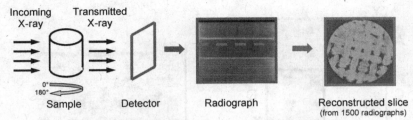

Figure 3: Schematic description of the X-ray tomography technique

In our experiment, images have been obtained by using the X-ray synchrotron source provided by ESRF on the ID19 beamline. Because of its parallel and monochromatic beam, synchrotron radiation tomography induces fewer artefacts than conventional X-ray tube tomography[2]. The energy of the beam was fixed to 40keV so that the average and minimum transmission were 29% and 21% respectively. The number of radiographs used for the reconstruction of the image is 1500 (that corresponds to an angular step of 0.12°). The detector, the FReLoN 2048x1700pixels camera of the ID19 beamline, with a pixel size of 5.02µm2, allows a voxel size of 5.02^3µm^3 (the terms "pixel" and "voxel" are respectively used for 2D and 3D images) for a sample size as large as 10.28mm (diameter) x 8.53 (height). The specimens used for this characterization were 10mm (diameter) x 3mm (height) disk of SiC/SiC composite. On Figure 3, a stack of three specimens can be observed on the radiograph. The present paper focuses on the characterization of one of these specimens. The choice of the resolution used for this characterization has been done in order to characterize the macro-porosity on a sufficiently large sample so that only the biggest pores of the micro-porosity are detected (3D characterizations of the micro-porosity is under progress with a higher resolution but on a smaller sample).

3 – ANALYSIS OF THE MACRO-POROSITY DISTRIBUTION

3.1 – Experimental results

3.1.1 – Region of interest and threshold operation

In order to avoid boundary effects, the following analysis is restricted to a region of interest extracted from the whole 3D image. The region of interest contains 1278x1278x660 voxels that corresponds to a volume of 6.4x6.4x3.3mm^3, which is assumed to be large enough to be representative of the composite's heterogeneity. Figure 4 presents a Z-slice (*i.e.* a 2D image parallel to the plane of the composite) of the region of interest. It can be seen that the size of this region includes 4 typical patterns of the plain weave architecture. Sub-regions marked as A, B, C and D are also defined on Figure 4, each region has the size of a typical pattern. An example of an X-slice (*i.e.* a 2D image perpendicular to the plane of the composite) is displayed on Figure 5 where the 13 plies can be observed.

In order to evaluate the distribution of the porosity, the voxels have to be separated into two phases: the black phase corresponding to the porosity and the white phase to the SiC/SiC material. The threshold used to separate the voxels is given by the local minimum observed on the grey level histogram of the 3D image and an example of this threshold operation can be observed on Figure 6.

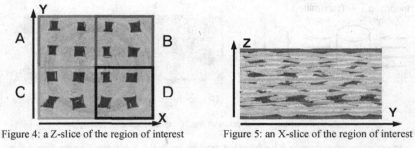

Figure 4: a Z-slice of the region of interest   Figure 5: an X-slice of the region of interest

Figure 6: definition of the threshold from the grey level histogram of the 3D image and example of the result of the threshold operation

3.1.2 – Porosity analysis

As mentioned in section 2.2, the tomographic porosity characterization includes the macro-pores and the biggest pores of the micro-porosity but from the observation of Figures 4, 5 and 6, this last contribution can be neglected. If the mean volume fraction of porosity is an important parameter, the main interest of such 3D characterizations is to study the spatial distribution of the porosity within the composite. The following results concern the evolution of the surface fraction of porosity in the different directions of the composite. The surface fraction of porosity is defined as the number of black voxels divided by the total number of voxels within a slice.

3.1.2.1 – Distribution of the porosity in the Z-direction

Figure 7a shows the surface fraction of porosity within XY-planes (*i.e.* the plane of the composite) as a function of the Z direction (*i.e.* the out-of-plane direction) for the whole region of interest. It can be observed that the surface fraction of porosity exhibits important fluctuations with well-defined maxima and minima. The value of the minima is quite constant (between 12% and 25%) whereas the values of the maxima are quite heterogeneous (between 22% and 48%). This conclusion is emphasized if we consider the sub-regions A, B, C and D separately (Figure 7b). Similar fluctuations of the porosity have already been shown on CVI SiC/SiC composites[3]. Moreover, it can be shown that these fluctuations are associated to the stacking of the plies within the composites. The minima are associated to the median plane *within* a ply: a typical Z-slice corresponding to a minimum can be observed on Figure 8b (the Z-position is marked as a square symbol on Figure 7a). On the other hand, the maxima can be associated to the median plane *between* two plies. Two examples of a Z-slice corresponding to a maximum can be observed on Figure 8a and 8c (the Z-positions are marked, on Figure 7a, as a circle and a triangle symbol respectively). Figures 8a and 8c reveal the existence of two different plies; their position in the XY-plane is represented by colored grids. On Figure 8a, that corresponds to a low maximum, the shift between the positions of the plies is important in the two directions (X and Y) whereas on Figure 8c, that corresponds to a high maximum, the shift between the positions of the plies is null in the vertical direction (Y). This effect of the relative position of the plies on the porosity distribution will be evidenced in section 3.2.

Finally, it seems important to point that a surface fraction of porosity of 60% on a surface of 3.2x3.2mm^2 (on sub-region A, Figure 9b) or of 48% on a surface of 6.4x6.4 mm^2 (Figure 9a) will have a significant effect on the out-of-plane thermo-mechanical properties of the composite.

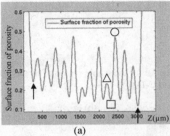 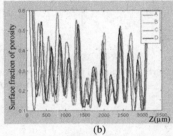

(a)   (b)

Figure 7: Surface fraction of porosity as a function of Z (μm): for the whole region of interest (a), for each sub-region A, B, C and D (b)

(a)   (b)   (c)

Figure 8: Z-slices corresponding to the triangle (a), square (b) and circle (c) symbols on Figure 7a. The positions of the different plies are represented by the colored grids.

3.1.2.2 – Distribution of the porosity in the X and Y-directions

On Figures 9 and 10 show the evolutions of the surface fraction of porosity in the X- and Y-direction respectively, for the whole region of interest (Figure 9a and 10a) or for the corresponding sub-regions (Figures 9b and 10b). In order to avoid boundary effects due to the free surfaces of the composite (Figure 5), the region of interest is limited in the Z-direction to the domain comprised between the two arrows on Figure 7a.

From the observation of these evolutions, the following points can be emphasized:
- the amplitude of the fluctuations in the X- and Y-direction is less important than in the Z-direction,
- in both directions, the surface fraction of porosity exhibits a quasi-periodic evolution and the corresponding period is comparable with the periodicity of the weave pattern (around 3mm). This result suggests that the 13 plies of the composite must have a preferential position in the XY plane, even if fluctuations around this position exist, as mentioned in the previous

section (this first result shows that a deeper analysis of the position of the plies has still to be performed).

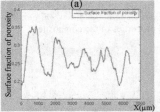 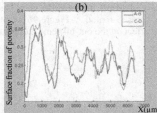

Figure 9: Surface fraction of porosity as a function of X (μm): for the whole region of interest (a), for each sub-region A-B and C-D (b)

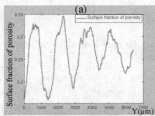 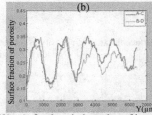

Figure 10: Surface fraction of porosity as a function of Y(μm): for the whole region of interest (a), for each sub-region A-C and B-D (b)

3.1.2.3 – Volume fraction of porosity

Finally, an average description of the porosity is given by the volume fraction of porosity which is equal, for the region of interest (limited in the Z-direction as in section 3.1.2.2), to 26.4%. The fluctuations of the volume fraction of porosity for the different sub-regions A, B, C and D, (25.7%, 25%, 29.2% and 25.9% respectively) are quite low so that the size of these sub-regions can be considered as large enough to be representative of the volume fraction of porosity within the composite. The value of 26.4% is consistent with the value of 27.5% obtained with a classical density measurement assuming that the density of SiC is 3.1 for both the fibres and the matrix.

3.2 – Effect of the stacking of the plies

3.2.1 – Evidence of the effect of the stacking of the plies on the porosity distribution

In order to discuss the distribution of the porosity as observed in the Z-direction (section 3.1.2.1), a geometrical model of the composite has been developed to evaluate the volume fraction of porosity as a function of the stacking of the plies.

As a first case, all the plies are considered to be exactly one above the other. As a consequence, the material can be represented by a quarter of a periodic unit-cell (Figure 11a). The cross section of the yarns is assumed to be an ellipse with an additional thin layer that represents the last deposition of SiC matrix (as observed on Figure 1), the neutral fibre of the yarn is defined by a spline curve, and all the geometrical parameters are consistent with the experimental observations. From this geometry, the volume fraction of porosity can be evaluated to 41%, which is not in agreement with experimental measurements. Thus, this configuration is not representative of the "real" material.

 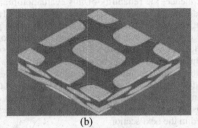

(a)   (b)

Figure 11: Geometry of a two volume elements considering: a unique ply (a), three successive plies with a relative position described on Figure 12 (b). Volume (a) is a quarter of a periodic volume element, volume (b) is a periodic volume element.

As a second case, 3 plies are considered in a periodic volume element. Figure 12 defines the vector that gives the relative positions, in the XY-plane, between the "mean" ply and the lower and upper plies. If this vector is null, the three successive plies are exactly one above the other and the configuration is exactly the same as the previous one. If this is not the case, the Z-position of the lower and upper planes have to be optimized in order to ensure a contact between the yarns of the successive plies. As a consequence, the compactness of the composite is increased, and the volume fraction of porosity is decreased. The compactness is maximized if the vector of relative positions is equal to $(0.25*p, 0.25*p)$ as presented on Figure 12, p being the weave pattern periodicity in the XY-plane. The corresponding microstructure is displayed on Figure 11b. The volume fraction of porosity evaluated on this microstructure is 21% which is in a better agreement with experimental results, compared to the previous case.

These two simple cases are clearly demonstrating the dependence of the local volume fraction of porosity as a function of the stacking of the plies. This dependence can explain the heterogeneity of the maxima observed on the evolution of the surface fraction of porosity as a function of Z (Figure 7).

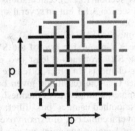

Figure 12: Relative position of the different plies

3.2.2 – Consequences on the micro-mechanical modelling

These results have an important consequence on the numerical evaluation of the effective properties of the composite from calculations performed on a Volume Element (VE). Such methods are now quite common[4,5,6,7...]: a 3D volume element is assumed to be representative of the microstructure of the composite (i.e. the VE is called a Representative Volume Element, RVE) and finite-element calculations are performed, generally with periodic boundary conditions, in order to evaluate the "effective" behaviour (*i.e.* the behaviour that links the average stress to the average strain on the RVE). As a consequence, the reliability of this evaluation mainly lies on the definition of a "representative" VE. As discussed above, a VE such as Figure 11a, generally assumed to be an RVE, is not "representative" of the variability of the stacking of the plies and, consequently, of the volume fraction of porosity observed within the composites. In order to account for this variability, one solution is to perform calculations on a larger VE that is really representative of the heterogeneity of the stacking of the plies. But such an RVE can be very large and results in prohibitive time calculations. An alternative method is to use a large number of smaller Volume Elements containing at least 2 or three successive plies, called Statistical Volume Elements (SVE), so that the variability of the stacking of the plies can be taken into account. The "apparent" behaviour is evaluated on each SVE, and the "effective" behaviour is estimated from the statistical average of these "apparent" behaviours. This last method will be described in the next section.

4 – MODELLING OF THE ELASTIC BEHAVIOUR

The purpose of this section is to deduce the influence of the macro-porosity on the "effective" elastic behaviour of the composite directly from the tomographic observation of the real microstructure. In the following, $\underline{a}$ denotes a vector, $\underline{\underline{a}}$ a second order tensor and A a fourth order tensor and the following notations (with the classical conventions on the summation on the repeated index[8]) are used:

$$\underline{a}.\underline{b} = a_{ij}b_j$$
$$A : \underline{\underline{b}} = A_{ijkl}b_{kl}$$

The overbar is used to denote the spatial average on the volume element Ω, with $\underline{x}$ the position of a point within Ω and V the volume of Ω:

$$\bar{a} = \frac{1}{V} \int_{\Omega} a(\underline{x})dV$$

4.1 – Methodology

4.1.1 – Definition of the "Statistical Volume Elements"

As discussed in the previous section (3.2.2), calculations are not directly performed on the whole region of interest described in section 3.1.1 but on several smaller regions that will be mentioned as Statistical Volume Elements (SVE).

The region of interest is divided into 24 SVE (Figure 13) so that each SVE has the following dimensions 3.2x3.2x0.45mm^3. In the XY-plane, the size of the SVE has the size of a typical weave pattern, and in the Z-direction, each SVE contains a least two plies. The finite element mesh is a regular mesh of linear 8-nodes cubic elements. In order to reduce the number of degrees of freedom, the refinement of the mesh is coarser than the resolution given by the tomographic image. For each SVE, the procedure for defining the material properties of the mesh comprises the following steps:

- the tomographic data are smoothed using a "box" filter: each voxel of the new image has the value of the mean value over a cubic box of $n \times n \times n$ voxels,
- the image is then reduced by keeping one point out of n in the different directions,
- the threshold procedure described in section 3.1.1 is then used to distinguish the porosity from the material, and a mesh of cubic elements is built with a direct correspondence between a voxel and a cubic element,
- finally, isotropic elastic properties are affected to each element: a value of 400GPa is assigned to the Young modulus of the "material" elements. In order to apply boundary

conditions (especially periodic BC), material properties with a very low stiffness (400GPa x10^{-6}) are artificially assigned to the "porosity" elements. The Poisson coefficient is 0.3.

In this work, the effect of the anisotropy of the tow (induced by the micro-porosity[1]) is not taken into account. As the fibre/matrix interface consists in a very thin layer (~100nm) of pyrocarbon, its volume fraction is less than 1%. Thus, its effect on the elastic behaviour of the composite can be reasonably neglected.

In order to evaluate the mesh sensitivity, three different refinements (Figure 14), given by the size of the "box" filter n (see above), are used: for a value of 9, the number of elements is ~55000 (71x71x11) and the corresponding size of each element is ~45x45x45μm^3; for a value of 7, the number of elements is ~110000 (92x92x13) and the corresponding size of each element is 35x35x35μm^3; for a value of 5, the number of elements is ~311000 (128x128x19) and the corresponding size of each element is ~25x25x25μm^3

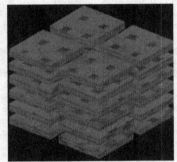

Figure 13: definition of the 24 "Statistical Volume Elements"

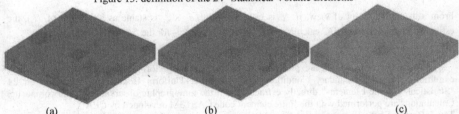

(a)   (b)   (c)

Figure 14: the three mesh refinements for an identical SVE: 71x71x11 (~55000) elements (a), 92x92x13 (~110000) elements (b), 128x128x19 (~311000) elements (c).

4.1.2 – Estimation of the "apparent" behaviour

The "apparent" stiffness tensor is defined as the fourth order tensor which links the average strain to the average stress (equation 1) obtained for 6 independent loadings applied on the SVE. The 6 loadings are said independent if the 6 resulting average stresses (or strains) are linearly independent.

$$\overline{\underline{\sigma}^I} = \widetilde{K}^{SVE} : \overline{\underline{\varepsilon}^I} \qquad\qquad I = 1..6 \qquad\qquad (1)$$

As a consequence, the value of the "apparent" stiffness tensor depends on the choice of the loadings which is itself associated to the choice of the boundary conditions used to perform the calculations. Classically, three different kinds of boundary conditions are used[9]:

- For Kinematic Uniform Boundary Conditions (KUBC), the displacement is imposed on the boundary (equation 2), with $\underline{\underline{\varepsilon}}^0$ a symmetric tensor:

$$\underline{u}(\underline{x}) = \underline{\underline{\varepsilon}}^0 . \underline{x} \qquad\qquad \forall \underline{x} \in \partial\Omega \qquad\qquad (2)$$

- For static uniform boundary conditions (SUBC), the normal stress is imposed on the boundary (equation 3).

$$\underline{t}(\underline{x}) = \underline{\underline{\sigma}}^0 . \underline{n} \qquad\qquad \forall \underline{x} \in \partial\Omega \qquad\qquad (3)$$

- Finally, for Periodic Boundary Conditions (PBC) the displacements of opposite points within opposite surfaces have to satisfy the periodicity condition given in equations (4).

$$\underline{u}(\underline{x}+\underline{h}) = \underline{\underline{\varepsilon}}^0 . \underline{h} + \underline{u}(\underline{x}) \qquad\qquad \forall \underline{x} \in \partial\Omega \qquad\qquad (4)$$

The "apparent" stiffness tensors, $\tilde{K}_{KUBC}^{SVE}$, $\tilde{K}_{PBC}^{SVE}$ and $\tilde{K}_{SUBC}^{SVE}$, obtained with these 3 kinds of boundary conditions are 3 different estimate of the "apparent" elastic behaviour. If the size of the SVE was large enough (*i.e.* if the size of the SVE was equal to the size of the RVE) these 3 estimates would be equivalent: the effect of the boundary conditions could be neglected and the "apparent" behaviour would be equal to the "effective" behaviour.

In our case, as the "porosity" (represented in our simulations by a very soft material) is connected to the boundary, the lower bound given by the SUBC is very low so that this lower bound is not significant. Therefore, in the next section, only the results obtained with KUBC and PBC will be presented.

4.1.3 – Estimation of the "effective" behaviour

As previously mentioned, the "effective" behaviour can be obtained from calculations performed on a RVE. Because of numerical limitations, calculations can only be performed on SVE and the "effective" behaviour can be estimated from a statistical average on the "apparent" behaviours evaluated on each SVE[9] (equation 5).

$$K_{BC}^{eff} = \left\langle \tilde{K}_{BC}^{SVE} \right\rangle = \frac{1}{N} \sum_{SVE} \tilde{K}_{BC}^{SVE} \qquad\qquad (5)$$

From a theoretical point of view, if N is large enough (*i.e.* K_{BC}^{eff} is stable as a function of N), the estimate obtained from KUBC calculations is an upper bound[10,11] for the "effective" behaviour.

4.2 – Numerical results

In this section, results are presented from calculations performed with 2 types of boundary conditions (Periodic Boundary Conditions and Kinematic Uniform Boundary Conditions) on 24 "Statistical Volume Elements" directly extracted from the tomographic observation of the composite. Calculations are performed with the finite element code CAST3M developed by CEA[12].

The "effective" behaviour estimated from PBC is assumed to be the most reliable estimation and the results obtained with Kinematic Uniform Boundary Condition are presented in order to discuss the size of the statistical volume elements. The mesh sensitivity is also discussed from the "apparent" behaviour estimated with the three mesh refinements (Figure 14) on two SVE. Finally, the fluctuations of the "apparent" behaviours evaluated on the 24 SVE are detailed in order to discuss the notion of "representative" volume element.

4.2.1 – Estimation of the "effective" behaviour

4.2.1.1 – PBC with the mesh refinement (b) (Figure 14)

The numerical results obtained with Periodic Boundary Conditions and with the refinement mesh (b) (Figure 14) are given on equation 6. It can be observed that the symmetry of the tensor is not so far from the expected quadratic symmetry (the X- and Y-directions are expected to be equivalent) and an approximation of this tensor with a quadratic symmetry is provided. This slight deviation from the quadratic symmetry can be explained by an insufficient number of SVE.

A common graphical representation of this result is shown on Figure 15 where the apparent Young modulus (the coefficient that relates the stress and strain for a uniaxial tensile test) evolves as a function of the tensile test direction, in the XY plane (Figure 15a) and in the XZ plane (Figure 15b). It can be observed that the anisotropy in the XY-plane is not very important (evolution between 222GPa for a tensile test direction at 45° from the X-direction and 250GPa in the X-direction) compared to the anisotropy in the XZ-plane (evolution between 122GPa for a tensile test direction in the Z-direction and 250GPa in the X-direction).

$$
K_{PBC}^{eff} = \begin{pmatrix}
289 & 83 & 58 & 1 & 2 & 0 \\
 & 278 & 53 & 1 & 1 & 0 \\
 & & 161 & 0 & 1 & 0 \\
 & & & 84 & 0 & 0 \\
 & SYM & & & 52 & 0 \\
 & & & & & 39
\end{pmatrix} \approx \begin{pmatrix}
284 & 83 & 56 & 0 & 0 & 0 \\
 & 284 & 56 & 0 & 0 & 0 \\
 & & 161 & 0 & 0 & 0 \\
 & & & 84 & 0 & 0 \\
 & SYM & & & 46 & 0 \\
 & & & & & 46
\end{pmatrix} \tag{6}
$$

4.2.1.2 – Sensitivity to the boundary conditions

The numerical results obtained with KUBC and mesh refinement (b) (Figure 14) are given on equation 7. As mentioned in sections 4.1.2 and 4.1.3, the effective stiffness tensor estimated with KUBC is stiffer than the one estimated with periodic boundary conditions. Moreover, it can be observed on Figure 15 that the anisotropy is underestimated compared to the anisotropy evaluated with PBC. However, the relatively small difference between the two types of boundary conditions allows to conclude that the size of the SVE used in these calculations is large enough (an important difference between KUBC and PBC results would have been the consequence of too small SVEs).

$$
K_{KUBC}^{eff} = \begin{pmatrix}
325 & 110 & 80 & 0 & 2 & 0 \\
 & 324 & 79 & 0 & 1 & 0 \\
 & & 198 & 0 & 1 & 0 \\
 & & & 103 & 0 & 0 \\
 & SYM & & & 75 & 0 \\
 & & & & & 75
\end{pmatrix} \approx \begin{pmatrix}
325 & 110 & 80 & 0 & 0 & 0 \\
 & 325 & 80 & 0 & 0 & 0 \\
 & & 198 & 0 & 0 & 0 \\
 & & & 103 & 0 & 0 \\
 & SYM & & & 75 & 0 \\
 & & & & & 75
\end{pmatrix} \tag{7}
$$

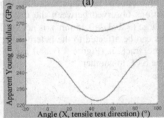
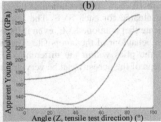

Figure 15: Evolution of the apparent Young modulus as a function of the tensile test direction, in the XY plane (a), in the XZ plane (b). Red = PBC; Blue=KUBC

4.2.1.3 – Sensitivity to the mesh refinement

In order to discuss the mesh sensitivity, the "apparent" stiffness tensors estimated with PBC and the three mesh refinements (Figure 14) are shown in table 1 for two different SVE. From table 1, it is clear that the mesh refinement (a) is not sufficient to estimate precisely the "apparent" behaviour of the SVE. This refinement is so coarse that the volume fraction of porosity is underestimated (table 1). On the other hand, the results obtained with the mesh refinement (b) are very close to the results obtained with the mesh refinement (c). However, calculations are still running to evaluate the "apparent" behaviour of each SVE with the mesh refinement (c). As observed in table 2, these calculations are time consuming.

Table 1: "Apparent" stiffness tensors and volume fraction of porosity estimated with different mesh refinements (Figure 14)

	Refinement (a)						Refinement (b)						Refinement (c)					
K_{PBC}^{SVE1}	311	98	70	1	−16	0	280	82	52	2	−16	0	281	81	48	1	−19	0
		319	72	1	−6	−1		290	55	2	−5	−1		289	50	1	−7	−1
			197	0	−6	0			154	0	−5	·0			139	0	−6	0
				95	1	−4				85	0	−4				86	0	−5
		SYM			55	0		SYM			39	0 ·		SYM			36	0
						65						51						48
f_v^{SVE1}	20.5%						23.4%						23.1%					
K_{PBC}^{SVE2}	291	88	71	−4	−5	1	264	73	56	−6	−5	2	259	70	49	−6	−5	1
		294	70	−4	−3	1		266	56	−5	−3	3		263	48	−5	−3	3
			209	0	−3	0			172	−1	−3	1			149	−1	−3	0
				84	0	−3				74	2	−3				74	2	−3
		SYM			62	1		SYM			49	0		SYM			44	0
						57						45						40
f_v^{SVE2}	22.5%						25.4%						25.3%					

Table 2: Estimation of the different time calculations for the evaluation of the "apparent" behaviour of one SVE (6 elastic calculations are required)

	KUBC(6 elastic calculations)	PBC (6 elastic calculations)
Mesh refinement (a)	0h15	0h30
Mesh refinement (b)	0h50	2h
Mesh refinement (c)	6h30	22h

4.2.2 – Fluctuation of the "apparent" behaviours

Figure 16 shows the evolution of the apparent Young modulus, as a function of the tensile test direction, evaluated for each SVE. The important fluctuations observed between the different SVE clearly indicate that a unique SVE, even if its size is quite large ($3.2 \times 3.2 \times 0.45 mm^3$), is not sufficient to describe the behaviour of the composite. These fluctuations can be affected to the heterogeneity of the stacking of the plies within the different SVE: the volume fraction (Figure 17) and especially the complex shape of the porosity can be very different from one SVE to another.

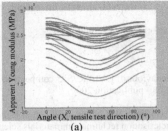
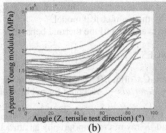

(a)   (b)

Figure 16: Evolution of the apparent Young modulus as a function of the tensile test direction, in the XY plane (a), in the XZ plane (b). Red = one curve per SVE; Blue = average behaviour

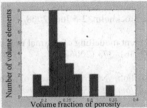

Figure 17: Histogram of the volume fraction of porosity for the 24 statistical volume elements

5 – CONCLUSION AND FUTURE PROSPECTS

A CVI SiC/SiC sample has been characterized by X-ray tomography (ESRF) and the distribution of its macro-porosity has been studied. The local volume fraction of porosity exhibits important fluctuations within the composite. A geometrical model has been developed to demonstrate that the stacking of the plies has a significant effect on the distribution of porosity. As a consequence, a micro-mechanical modelling based on calculations performed on a Volume Element (or unit-cell) that contains only one ply is not sufficient to account for the variability of the stacking of the plies. In order to take into account this variability, several volume elements have to be considered and each volume element must contain at least two successive plies. These volume elements are then called "statistical volume elements".

In order to evaluate the effect of the macro-porosity on the elastic behaviour, calculations have been performed on 24 "statistical volume element" (SVE) directly extracted from the X-ray tomography characterization. The "effective" behaviour of the composite has been evaluated from an average of the "apparent" behaviour estimated for each SVE with Periodic Boundary conditions. A quite strong anisotropy has been evidenced. Kinematic Uniform Boundary Conditions have also been used, and the reasonable difference between KUBC and PBC confirms that the size of the SVE is not "too small". Finally, it has also been confirmed that a unique SVE, in spite of its quite important size of $3.2 \times 3.2 \times 0.45 mm^3$, is not representative of the behaviour of the composite.

In a future work, the relative position between the successive plies will be deeply studied in order to distinguish the most represented configurations and to reproduce "statistical volume element" from the geometrical model proposed in section 3.2.

In this work, the anisotropy of the tow (induced by the micro-porosity[1]), has not been considered. Its effect on the behaviour of the composite will have to be evaluated. In order to account for this effect, the definition of local axis will have to be affected at each point of the tow. This operation is quite

complicated on tomographic images and it will be more convenient to work on volume elements generated by the geometrical model.
Finally, the evaluation of the thermal behaviour will be performed in a similar way.

REFERENCES

[1] L.Gélébart, C.Colin, Effects of porosity on the elastic behaviour of CVI SiC/SiC composites, *Journal of Nuclear Materials* – (Accepted for publication April 2008 - publication 2009)

[2] L.Salvo, P.Cloetens, E.Maire, S.Zabler, J.J.Blandin, J.Y.Buffière, W.Ludwig, E.Boller, D.Bellet, C.Josserond, X-ray micro-tomography an attractive characterisation technique in materials science, *Nuclear Instruments and Methods in Physics Research B*, **200**, 273–286 (2003)

[3] A.Morales-Rodríguez, P. Reynaud, G.Fantozzi, J. Adrien, E. Maire, 3D-porosity structural analysis of long - fibre - reinforced ceramic matrix composites using X-ray tomography, *13th European Conference on Composite Materials*, Stockholm, 2-5 June 2008

[4] J.K.Farooqi, M.A.Sheikh, Finite element modelling of thermal transport in ceramic matrix composites, *Computational Materials Science*, **37**, 361–373 (2006)

[5] E.J.Barbero, T.M.Damiani, J.Trovillion, Micromechanics of fabric reinforced composites with periodic microstructure, *International Journal of Solids and Structures*, **42**, 2489–2504 (2005)

[6] D.Scida, Z.Abouraa, M.L.Benzeggagh, E.Bocherens, A micromechanics model for 3D elasticity and failure of woven-fibre composite materials, *Composites Science and Technology*, **59**, 505-517 (1999)

[7] S.V.Lomov, A.V.Gusakov, G.Huysmans, A.Prodromou, I.Verpoest, Textile geometry preprocessor for meso-mechanical models of woven composites, *Composites Science and Technology*, **60**, 2083-2095 (2000)

[8] M.Bornert, T.Bretheau, P.Gilormini, Homogénéisation en Mécanique des Matériaux, Hermes, ISBN 2-7462-0199-2, (2001)

[9] T.Kanit, S.Forest, I.Galliet, V.Mounoury, D.Jeulin, Determination of the representative volume element for random composites: statistical and numerical approach, *International Journal of Solids and Structures*, **40**, 3647-3679 (2003)

[10] C.Huet, Application of variational concepts to size effects in elastic heterogeneous bodies, *Journal of the Mechanics and Physics of Solids*, **38**, 813-841 (1991)

[11] M. Ostoja-Starzewski, Material spatial randomness: from statistical to representative volume element, *Probabilistic Engineering Mechanics*, **21**, 112-132 (2006)

[12] CAST3M, finite element code developed by CEA, http://www-cast3m.cea.fr

MECHANICAL STRENGTH OF CTP TRIPLEX SIC FUEL CLAD TUBES AFTER IRRADIATION IN MIT RESEARCH REACTOR UNDER PWR COOLANT CONDITIONS

Herbert Feinroth
Ceramic Tubular Products LLC
Rockville, Maryland USA

Matthew Ales and Eric Barringer
Ceramic Tubular Products LLC

Gordon Kohse and David Carpenter
Massachusetts Institute of Technology
Cambridge, Massachusetts USA

Roger Jaramillo
Oak Ridge National Laboratory
Oak Ridge, Tennessee USA

ABSTRACT

An experiment was conducted in the MIT Research Reactor (MITR) to irradiate triplex silicon carbide fuel cladding tubes under typical Pressurized Water Reactor conditions. Measurements were made to determine the impact of exposure on strength and swelling. The SiC clad tubes were fabricated by Ceramic Tubular Products (CTP) with dimensions typical of 15 x 15 commercial PWR reactor fuel. The triplex tubes contain 3 layers, an inner monolithic SiC layer to maintain hermeticity, a central SiC/SiC composite layer to provide a graceful failure mode in the event of an accident, and an outer SiC environmental barrier layer. Clad tubes were exposed to 300 °C pressurized water containing boric acid, lithium hydroxide, and hydrogen overpressure, typical of PWRs. Thirty nine (39) specimens of various types were exposed to coolant, some within the neutron flux region and some outside the neutron flux region. Twenty seven (27) were removed for examination and test after 4 months exposure. Following examination, twenty specimens were reinserted for additional exposure, along with 19 new specimens. The 4 month specimens were weighed and measured at MIT, and some were shipped to Oak Ridge National Laboratory (ORNL) where they were mechanically tested for hoop strength using a polyurethane plug test apparatus. Results were compared with the pre-irradiation strength and dimensions. Some specimens retained their original strength after exposure, others with a less homogeneous monolith, lost strength.

INTRODUCTION

Gamma Engineering, one of the parent companies of Ceramic Tubular Products, has been developing multilayered silicon carbide clad tubes for use in commercial light water reactors since 1999. Reference (1) summarizes the initial development work performed under a DOE small business grant. This early work confirmed that a two layer silicon carbide clad tube, combining a dense inner monolith layer with a composite outer layer, is impermeable to fission gases during normal reactor operation, and exhibits a graceful failure mode during severe accidents. This latter behavior allows the fuel clad to retain the solid uranium oxide fuel and also to maintain coolability during severe accidents.

After 2001, Gamma continued to refine and test the multilayered concept. Several different fiber architectures, fiber types and matrix infiltration methods were examined in an effort to assure radiation resistance and to provide increased strength of the multilayered tube for both normal operation and accident conditions. Reference (2) summarizes early mechanical hoop strength tests of the un-irradiated duplex tubes using the plug testing approach developed by ORNL for the Mixed Oxide Fuel test program. Details of this hoop test method are described in reference (2).

In May, 2006, with support from both the DOE and Westinghouse, an irradiation test was initiated in the MIT research reactor (MITR) to examine the effects of exposure to neutron irradiation and hot PWR coolant on the behavior of multilayered silicon carbide fuel clad tubes. Tube diameter and thickness of the test specimens were almost typical of the 15 x 15 commercial PWR fuel assemblies now in use. Test specimens were about 2 inches long. After initial exposure of 39 specimens for a period of 4 months, about half the specimens were removed for examination and testing, and replaced with new specimens for continued irradiation. The removed specimens were allowed to cool, examined for dimensional changes, and then tested for hoop strength to determine the effect of irradiation on strength. This paper describes the multilayered clad specimens, the conditions of the exposure in the MITR, and the results of the strength testing and dimensional examination. It also describes more recent measurement of dimensional changes after exposure for an additional 8 months. Additional mechanical tests (hoop strength) after the additional 8 months are planned in the future.

SAMPLE PREPARATION

Because funding was limited, post irradiation hoop strength testing was performed on only five tube specimens. Each specimen was originally 1.9 inch in length during irradiation, and about 1.4 inch in length during post irradiation mechanical testing. The specimens were selected for this irradiation exposure and post irradiation examination on the basis of having a variety of different monolith characteristics, fiber suppliers and outer environmental barrier coating, in order to learn which of these characteristics offered the best performance during exposure.

During irradiation, the specimens were 1.9 inches long, and were exposed to coolant both inside and out, and at the cut ends, although coolant flow velocities were much lower on the inner and exposed end surfaces. After exposure, a ¼ inch ring was removed from each end for corrosion studies. Hence, the post irradiation mechanical testing was performed on 1.4 inch long specimens. Except for specimen H2-5, all specimens were exposed to neutron irradiation. Specimen H2-5 was exposed to coolant but not to in-core neutron flux.

Characteristics of the five exposed and tested specimens are presented in Table I:

Table I – Test Specimen Characteristics

Specimen Identity	Monolith	Fiber	Barrier coating	Core Location	% of peak fluence
H2-1	Coorstek	Sylram iBN	CVI cap	Tier 3	95
H2-5	Coorstek	Sylram iBN	CVI cap	Out of core	0
F1-1	Coorstek	HiNic-S	CVI cap	Tier 3	95
D1-2	TREX	Sylram iBN	Machine + TREX cvd	Tier 4	100
B1-2	TREX	HiNic-S	CVI cap	Tier 2	90

Common characteristics – All specimens incorporated a beta phase chemical vapor deposition (CVD) monolith of about 15 mils thickness and 0.350 – 0.354 inch ID, a fiber wound architecture developed by Ceramic Tubular Products using a unique winding device, a carbon interface coating on the fibers, and a composite layer made using chemical vapor infiltration (CVI). As noted in Table 1, four of the specimens incorporated an outer environmental barrier coating (EBC) which was merely an extension of the CVI composite layer after the infiltration "capped off", whereas the fifth specimen, D1-2, had a unique EBC in which the "capped off" CVI layer was machined smooth, and an additional SiC layer was deposited by TREX using CVD. Both types of fibers are stoichiometric beta phase fibers. The Sylramic iBN fiber also has a thin interface layer of in-situ diffused boron carbide.

IRRADIATION CONDITIONS IN THE MIT TEST REACTOR

Although the 5 MW MIT research reactor operates at near atmospheric pressure and temperature, a special titanium autoclave inserted in the reactor, and connected to a closed loop outside the reactor, provides inlet water at 300 °C and 1500 psi at a mass flow of 0.25 kg/sec. Oxygen is removed from the circulating water by bubbling hydrogen through the coolant makeup tank and a catalytic recombiner. The heated portion of the loop is insulated by a CO_2 gas gap, assuring a temperature drop of not more than 20 °C over the length of the in-core section. Thus, this closed loop simulates the actual coolant conditions in a typical PWR commercial reactor.

The design of the titanium autoclave also allows for a portion of the test specimens to operate in the PWR coolant but outside of the high neutron flux core area. This allows separating the effects of coolant corrosion only, from the combined effects of coolant exposure and neutron irradiation.

Figure 1 presents the neutron flux profile for the MIT reactor. The MIT staff has estimated an average neutron fluence (> 1 MeV) during the 4 month test of 0.5×10^{21} n/cm^2 based on neutron activation of the titanium loop materials. Based on the flux profile in Figure 1, we calculate that the central tiers of this experiment (tiers 2, 3, 4 and 5) received about 0.63×10^{21} neutrons/cm^2.

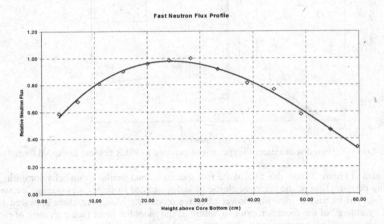

Fast Neutron Flux Profile

Figure 1 - Neutron Flux Profile in MITR-II Reactor

With respect to fast neutron damage, one displacement per atom (dpa) for silicon carbide is achieved with a fast neutron fluence (> 1 MeV) of about 1×10^{21}. Thus during this experiment, the four irradiated specimens that were mechanically tested received about 0.63 dpa exposure. For comparison purposes, the clad in a fully burned LWR fuel rod (50 gwd/t) receives about 10 dpa fast neutron exposure.

MECHANICAL TESTING AT ORNL

ORNL successfully completed the room-temperature hoop strength tests for the five 1.40" specimens supplied by MIT. Due to the high levels of contamination for these specimens, testing was performed in a hot cell located within Building 3525.

Figure 2 shows the load vs. ram displacement data for the five specimens. The specimen load is the ram force minus the force required to compress the plug without a specimen. The x axis has been normalized such that the origin corresponds with the onset of specimen loading.

This figure indicates that H2-1 and H2-5 had the largest load bearing capability of 850 pounds, and F1-1 had a bit lower load capability of about 700 pounds. Specimens D1-2, and B1-2 both had about half the load capability of the H series, 400 to 450 pounds. As noted below, we ascribe this decrease in capability to structural defects (laminations and porosity) in the monolith layer.

The general failure behavior for all of the specimens was similar to that previously observed for as-fabricated specimens. That is, the specimens were loaded until the monolith layer failed, at which point all load was transferred to the composite/EBC layers. For each specimen, load was maintained as ram displacement was continued.

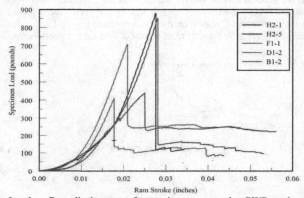

Figure 2. Load vs. Ram displacement for specimens exposed to PWR coolant in the MIT reactor.

The data in Figure 2 show that four of the five specimens had similar strain behavior during the initial loading period. That is, the effective elastic modulus (related to slope) was similar. However, B1-2 was observed to have a somewhat lower effective modulus. Moreover, there appeared to be some initial "yielding" of the specimen prior to failure of the monolith layer (see curvature of plot in Figure 2). This behavior is inconsistent with that observed for the other SiC specimens, including the as-fabricated specimens. The reasons for this difference in strain behavior are not known at this time.

COMPARISON WITH AS-FABRICATED SPECIMEN BEHAVIOR

The room-temperature hoop strength data for the five exposed specimens (peak load at failure) are compared to test results obtained for as-fabricated specimens in Figure 3. The hoop strength for F1-1, H2-1 and H2-5 where within the range measured for the as-fabricated specimens. The as fabricated test results comprised a larger data base and the average strength and standard deviation are as follows:

- F-Series Specimens – 4 data points; average strength = 808 lbs; SD = 85 lbs
- H-Series Specimens – 3 data points; average strength = 881 lbs; SD = 8 lbs

There appears to be a substantial reduction in strength during irradiation for the two specimens (B1-2 and D1-2) fabricated with the TREX monolith. It should be noted that the number of specimens fabricated with the TREX monolith that have been tested is somewhat limited, and caution must therefore be exercised when interpreting these results.

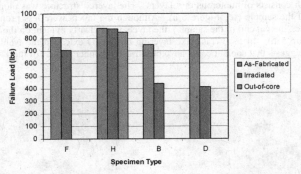

Figure 3. Comparison of room-temperature hoop strength for as-fabricated specimens and for specimens exposed to PWR coolant in the MIT Research Reactor.

RESULTS OF SEM EXAMINATIONS

Test specimens were examined visually and by SEM after 4, 8 and 12 months exposure. Figure 4 depicts an F series specimen after 12 months exposure and prior to removal of the ¼ inch end ring for final weight change measurements. The small triangular piece of composite missing at the top end is thought to have resulted from a combination of the mechanical fixture used to hold the specimens in place, and possible corrosion of the central layer starting from the exposed cut ends. The figure also illustrates the winding pattern, or fiber architecture used in CTPs triplex cladding design. Although not visible on the photograph, there is a 5 mil overcoat of SiC that follows the pattern of the winding, and that provides protection from coolant corrosion effects.

Figure 4 - Specimen F1-4 after 12 months exposure in MIT reactor

To provide some insight into the strength test results, SEM examination was performed on small ring sections of several of these specimens that were cut from the ends by MIT prior to conducting the hoop strength tests at ORNL. Figure 5 shows a high magnification (1,000 x) image of the inner surface of the TREX monolith on a B series specimen. The section is an axial cut of the ¼ inch ring cut from the end of the specimen, with the right side of the image being the inner surface of the monolith tube exposed to coolant. Examination of the polished cross section showed that the TREX monolith consists of numerous thin layers. The layered structure was more prominent on the exposed end of the sample (see Figure 5 a). Although the layers were less prominent ¼" from the exposed end (i.e., the cut end of the sample), the layers were clearly evident on the inner surface of the monolith (Figure 5 b). In fact, crack-like layering on the ID extended the whole ¼" length of the sample. This suggests that coolant could have penetrated the inner surface of the TREX monolith.

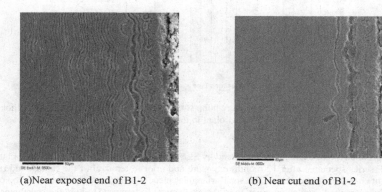

(a)Near exposed end of B1-2   (b) Near cut end of B1-2

Figure 5 SEM's of Monolith Portion of Sample B1-2 (TREX monolith) after 4 month exposure

It is speculated that the reason for the lower strength of the B and D specimens after irradiation is at least partly due to the porosity and laminations in the monolith and the possible effects of irradiation on the behavior of the duplex structure.   Further investigation revealed that the supplier used a "research" type reactor in preparing the CVD material, with time variant deposition of higher carbon non-stoichiometric SiC. The layers of non-stoichiometric higher carbon SiC had accelerated corrosion leading to the observed reduction in strength.   For later development work, the supplier utilized an improved   deposition   reactor,   and   conventional   CVD   process   parameters,   leading   to   more homogeneous, stoichiometric, dense material.

DIMENSIONAL CHANGES AND RADIATION INDUCED SWELLING
After both the 4 month exposure, and the additional 8 month exposure, post-irradiation dimensions were measured for some triplex tube specimens. The data are provided in Table II. The OD measurements have high variability because of the fiber architecture, and therefore are not a reliable measure of radial swelling. Axial growth measurements are more accurate.

Table II. Dimensional changes for irradiated triplex tube specimens

| Sample | Exposure Time (mo) | Pre-irradiation | | Post-irradiation | | Irrad Δlength | | Irrad ΔOD |
		Length (in)	Avg OD (in)	Length (in)	Avg OD (in)	(inches)	(percent)	(inches)
F4-1	8	1.904	0.414	1.915	0.416	0.011	0.58%	0.002
F2-3	8	1.904	0.411	1.913	0.416	0.009	0.47%	0.005
H1-4	8	1.896	0.421	1.907	0.424	0.011	0.58%	0.004
H1-2	8	1.906	0.421	1.918	0.424	0.012	0.63%	0.003
B1-4	12	1.909	0.419	1.920	0.421	0.011	0.58%	0.002
H3-4	12	1.893	0.419	1.907	0.421	0.014	0.74%	0.002
F1-4	12	1.871	0.410	1.883	0.412	0.012	0.64%	0.002

The average length growth for the specimens irradiated for 8 months (average exposure of about 1.3 dpa) and 12 months (average exposure of about 1.9 dpa) are summarized in Table III.

Table III – Average length growth of SiC triplex cladding after MITR irradiation

Exposure	Average	Std. Deviation
8 Month	0.57%	0.066%
12 Month	0.65%	0.082%

These results compare well with the data reported by Snead, et al., showing saturated volumetric swelling of about 2% (equivalent to 0.67% axial swelling) at about 300 °C. The Snead compilation (redrawn from reference 3) is shown in figure 6.

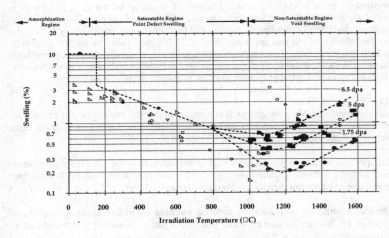

Figure 6 - Volumetric Swelling for SiC (redrawn from reference 3)

ANALYSIS AND EVALUATION OF MECHANICAL TEST RESULTS

CVD beta phase silicon carbide is reported by the suppliers (Coorstek and TREX) to have a flexural strength of 450 to 468 MPa (66,000 to 69,000 psi.) The use of flexural strength as the failure strength for hoop loading was validated by testing samples of monolith. Using the thin walled cylinder formula (stress = pressure x radius/thickness), we calculated what the stress in the monolith layer would be (in the five specimens tested) if all of the hoop load was taken by the monolith layer, with none of the hoop load being shared with the composite layer or EBC prior to failure of the monolith. The results are shown in Table IV.

Table IV Internal Pressure and Stress in Monolith at Failure Assuming No Load Sharing

	Type	Failure load lb	Internal pressure psi	Stress at failure psi	Possible Composite/EBC load share
F1-1	Coorstek HiNicalon	708	7516	86,760	21.6%
H2-1	Coorstek Sylramic	880	9359	107,940	37%
H2-5	Coorstek Sylramic	851	9009	104,144	34.7%
B1-2	TREX HiNicalon	439	4542	53,111	NA
D1-2	TREX Syl w/ovct	411	4235	50,675	NA

The results show that the three specimens with Coorstek monolith exhibited an apparent strength which is 21.6 to 37% higher than the reported flexural strength of the monolith alone.   Since we know that the actual stress in the monolith will not exceed the basic material capability of about 68,000 psi, the results of these tests show that the internal pressure load was actually shared amongst the three layers prior to monolith failure.   What is revealed is that the combination of the tightly wound composite layer, and the outer environmental barrier layer, have reinforced the monolith layer, and these two added layers together share from 21.6 to 37% of the total hoop load applied in the test.

This result is quite important to the application of this cladding to PWR cladding. In effect, it demonstrates the capability of the cladding to take 3 to 4 times more fission gas pressure during reactor operation, as compared to zircaloy cladding which because of its creep behavior, is limited to a maximum pressure of about 2000 psi. It is this ability to take high internal pressure that exemplifies the value of this ceramic cladding to extending the burnup and lifetime of light water reactor fuel.

With regard to the behavior of the two specimens having TREX monoliths, we cannot determine whether there was load sharing in these triplex tubes because, as stated previously, the presence of high porosity and laminations in these tubes seems to have decreased strength overall, and it is not possible to separate the effects of reduced material properties from the possible effects of load sharing.

SUMMARY AND CONCLUSIONS

Irradiation testing of triplex cladding specimens is continuing in the MIT reactor, and is planned to begin with contained fuel in the HFIR reactor in 2009.   The results of initial post irradiation tests described in this report indicate that:

-   Triplex clad specimens made with good quality CVD monoliths have the same hoop strength as un-irradiated samples within 2 standard deviations, after exposure to typical PWR operating conditions including fast neutron irradiation averaging about 0.63 dpa.

-   Average length growth due to irradiation induced swelling of triplex cladding specimens exposed to about 1.9 dpa fast neutron exposure was 0.65% and is consistent with the values in previous experiments.

-   The addition of an outer composite layer with a unique fiber architecture, and an outer environmental barrier layer, to a monolith tube, can provide reinforcement to the monolith layer on the order of 21.6 to 37%, thus providing additional pressure retention capability to triplex tubes during reactor operation. This increases the capability for SiC triplex cladding (as compared to zircaloy cladding) to safely allow higher fission gas release, as required for high burnup, high power rated, nuclear fuel.

ACKNOWLEDGEMENTS

The authors wish to acknowledge the contributions made by members of several institutions who contributed to this research, including Yakov Ostrovsky and Mujid Kazimi from MIT, Edgar Lara Curzio and the staff of the High Temperature Materials Lab at Oak Ridge National Laboratory, Ed Lahoda from the Westinghouse Research and Development Division, and Madeline Feltus from the US Department of Energy, Office of Nuclear Energy.

REFERENCES
1) H. Feinroth, B. Hao, L Fehrenbacher, M. Patterson, "Progress in Developing an Impermeable High Temperature Ceramic Composite for Advanced Reactor Clad and Structural Applications" Paper 1176, International Conference on Advanced Nuclear Power Plants, Hollywood, Florida, 2002.
2) Denwood F. Ross, William R. Hendrick, "Strength Testing of Monolithic and Duplex Silicon Carbide Cylinders in Support of Use as Nuclear Fuel Cladding" The 30th International Conference on Advanced Ceramics & Composites (2006 ICACC) January, 2006 – Cocoa Beach, Florida
3) L. Snead, Y. Katoh, S. Connery, "Swelling of SiC at Intermediate and High Irradiation Temperatures" Journal of Nuclear Materials, **367-370** 677-684 (2007)

Mechanical Properties

BEHAVIORS OF SiC FIBERS AT HIGH TEMPERATURE

C. COLIN, V. FALANGA, L. GELEBART
CEA, Nuclear Energy Division, Nuclear Material Department, Section for Applied Metallurgy Research,
DEN/DMN/SRMA, CEA Saclay, 91 191 Gif-sur-Yvette Cedex,
FRANCE

ABSTRACT

On the one hand, considering the improvements of mechanical and thermal behaviours of the last generation of SiC fibers (Hi-Nicalon S, Tyranno SA3); on the other hand, regarding physical and chemical properties and stability under irradiation, SiC/SiC composites are potential candidates for nuclear applications in advanced fission and fusion reactors. CEA must characterize and optimize these composites before their uses in reactors. In order to study this material, CEA is developing a multi-scale approach by modelling from fibers to bulk composite specimen: fibres behaviours must be well known in first. Thus, CEA developed a specific tensile test device on single fibers at high temperature, named MecaSiC. Using this device, we have already characterized the thermoelastic and thermoelectric behaviours of SiC fibers. Additional results about the plastic properties at high temperatures were also obtained. Indeed, we performed tensile tests between 1200℃ up to 1700℃ to characterize this plastic behaviour. Some thermal annealing, up to 3 hours at 1700℃, had been also performed. Furthermore, we compare the mechanical behaviours with the thermal evolution of the electric resistivity of these SiC fibers. Soon, MecaSiC will be coupled to a new charged particle accelerator. Thus, in this configuration, we will be able to study in-situ irradiation effects on fibre behaviours, as swelling or creep for example.

INTRODUCTION

Owing to progress in the manufacturing of SiC fibers (Hi-Nicalon[TM] S, Tyranno[TM] SA3), the mechanical and thermal behaviours of SiC_f/SiC_m composites have been sharply improved. Besides, regarding their physical and chemical properties and their stability under irradiation, SiC/SiC composites are potential candidates for nuclear applications in advanced fission (Generation IV) and fusion reactors (ITER). For example, these composites are intended as fuel cladding in Gas cooled Fast Reactor (GFR). These applications imply a severe environment including high temperature, fast neutrons spectrum and surrounding coolant and fuels. That is why CEA launched a specific program to characterize and optimize SiC/SiC composites before their use in reactors. Among the numerous investigations aiming to evaluate this material, CEA is developing a multi-scale approach and modelling to forecast the behaviour of SiC_f/SiC_m; this requires very good knowledge of each constituent of the composite (fiber, interphase, matrix).

A previous study dealt with the thermo-electric and thermo-elastic properties of these SiC fibers and matrix [1]. Now, the purpose of this work is to determine the behaviours of SiC fibers at high temperature. More specifically, it concerns the plastic tensile properties of fibers.

59

EXPERIMENTS AND MATERIALS

CEA has developed in collaboration with LCTS[*] a specific device for tensile tests at high temperature on fibers, named MecaSiC. It is described more specifically elsewhere [1, 2]. A schematic diagram of this device is shown in figure 1. Tensile tests can be performed from room temperature up to 1800°C under secondary vacuum (residual pressure below 10^- ^{6}mbar) on SiC fibers or on microcomposites with a gauge length of 25mm. Specimens are bonded on graphite grips using C34 UCAR cement. The main novelty is found in the heating specimen: the high temperatures can be reached by applying an electric current to the fibers. As a consequence, the electrical conductivity can be easily estimated by measuring resistance and length of fibers. The cross-section of SiC fibers is assumed to be circular and the diameter (average of five measurements per fiber) is measured, before tests, using laser diffractometry. Sample strain is derived from grip displacement by using a compliance calibration technique. The load cell is in the vacuum chamber.

Two different SiC fibers are examined in this study: Hi-Nicalon[TM] Type-S (HNLS) and Tyranno[TM] SA3 (TSA3). Table 1 lists fibers properties provided by the manufacturers and various references [3-6].

The experimental procedure had been described in [1]:

- Fiber temperature measurement and estimation: it is shown that, with this heating method, the fiber temperature can be considered as uniform along and in the section of the fiber [7]. Temperature at the fiber surface can be measured by using a bichromatic pyrometer. In practice, only temperatures in the range 1200-2000°C are measured on smallest fibers (TSA3 is sometimes less than 7µm in diameter). Since tests are performed under secondary vacuum, it is assumed that heat all the supplied electric power P_{elec} is converted to thermal radiation of the fiber. So the specimen temperature T_{est} can be estimated by the radiation law $P_{elec} \propto \left(T_{est}^{\,4} - T_{RT}^{\,4}\right)$. At high temperature, typically over 1200°C, the estimated temperature, T_{est}, introduces an uncertainty of about 50°C.

- Determination of Young's modulus and tensile strength of fibers: tensile tests were carried out at a strain rate of 10^{-4} s^{-1}. The influence of strain rate is not considered in this paper. The Young's modulus is estimated between 50 and 350MPa. Young's modulus is the average of the slopes extracted from the three stress-strain curves. The tests were performed from room temperature up to 1700°C (estimated temperature T_{est}, or measured temperature T by pyrometer). The uncertainty on elasticity properties estimations is less than 5%.

- Measurement of electrical resistivity: during these tensile tests, the fiber electrical resistivity, ρ, was estimated during the different temperature steps by $\rho = R \cdot Sect/L$, where R is the resistance (R=U/I), Sect is the section area and L is the length of fiber, at the considered temperature. The uncertainty on fiber resistivity measurement is less than 5%.

RESULTS

This part will consist in a brief description of the obtained results. Indeed, a more detailed paper is in progress and will soon be submitted to publication: it will describe the results in details and will present an extended discussion. To obtain further details, the reader will be interested in this article. The aim of this study is to characterize the mechanical behaviour of the two kinds of SiC fibres at high temperature, using MecaSiC.

The previous study, [1], shown in particular that in the range 20-1800°C, the electrical resistivity of the two kinds of SiC fibers exhibits the same sigmoid trend in heating conditions. Then the resistivity decreases steeply up to 1200°C. At higher temperature, the resistivity decreases slower. The cooling curves lay below the heating curves. It corresponds to an increase in conductivity. Many authors have already related this point. It reflects changes in microstructure or chemical modifications at high temperature: the fibers could be covered by carbon [4, 8, 9]. Dicarlo, [10], evoked a volatilization phenomenon of silicon or of silicon oxides during experiments at high temperature under controlled atmosphere. Sauder had also put in evidence the same phenomenon on facies observations after creep test on HNLS or TSA3 fibers [9].

So the first step of this study enabled to identify and quantify the changes at high temperature. On the one hand, on the TSA3 fibres, microstructural observations showed the silicon gaseous volatilization after 2 hours exposure at high temperature under vacuum, leaving a porous carbon layer in the periphery of the fibre (figure 2). The volatilization kinetics has been determined. On the contrary, it was difficult to measure silicon volatilization on HNLS fibres. A comparison has been made between previous microstructural observations and evolutions of mechanical and electric properties according to temperature. The Young modulus and electrical resistivity were taken into consideration. The whole observations show after all a limited kinetic at the highest temperature. So during the short tensile test (less 10min), we disregarded this phenomenon, assuming that there is no important change in fibres during their heating and during their failure test.

Then, tensile tests were performed at high temperatures, between 1200 and 1700°C, in order to study the plastic behaviour of HNLS and TSA3 fibers at a strain rate of 10^{-4} s^{-1}. An example of the obtained stress-strain curves on TSA3 fibers is visible on figure 3. We performed the same kind of tests on HNLS fibers. With these data, we estimated the elastic-plastic transition temperature, and we measured the tensile strength at high temperature. For both fibers, measurements showed a decrease in rupture and yield stresses, and an increase in rupture strain depending on temperatures. We tried to compare our results with the literature, but in most of case the strength of fibers are measured after heat treatment, [4] for example. In spite of this experimental procedure, our results seem to be consistent with the previous studies: the absolute values are quite smaller but the decreases of properties occur at the same temperature.

CONCLUSIONS

CEA has developed a specific device for tensile tests at high temperature on fibers, named MecaSiC. Tensile tests can be performed from room temperature up to 1700°C under secondary vacuum (residual pressure below 10^{-6}mbar) on SiC fibers or on microcomposites with a gauge length of 25mm. Two different last generation SiC fibers are examined in this study: Hi-Nicalon™ Type-S (HNLS) and Tyranno™ SA3 (TSA3).

The following results have been displayed:

- A volatilization of silicon phenomenon occurs at high temperature under vacuum, mainly on TSA3 fibers. But considering the kinetics, it is assumed that this phenomenon is neglected during our experiments.

- The elastic-plastic transition temperatures were estimated with different tensile tests and we measured the tensile strength at high temperature.

- For both fibers, measurements showed a decrease in rupture and yield stresses, and an increase in rupture strain depending on temperatures.

- Our results seem to be consistent with the previous papers.

Soon, MecaSiC will be coupled to JANNUS, based in CEA Saclay. JANNUS (Joint Accelerators for Nano-Science and Nuclear Simulation) is a project designed to study the modification of materials using multiple ion beams and in-situ TEM observation. In this configuration, we will be able to study in-situ irradiation effects on fiber behaviours, as swelling or creep for example.

Table 1 - Main characteristics of- as-received fibers at room temperature

Fibers	HNLS	TSA3
Mean diameter (µm)	12	7.5
Strength (GPa)	2.6	2.7
Young's modulus (GPa)	360	390
Composition		
Oxygen (wt%)	0.2	< 0.5
C/Si	1.05	1.15-1.04

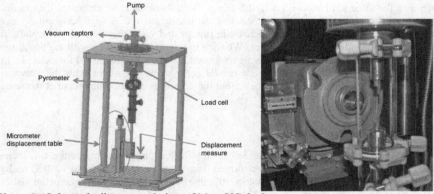

Figure 1 - Schematic diagram and view of MecaSiC device.

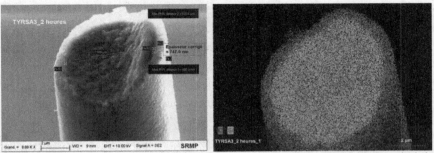

Figure 2 – Observations of a TSA3 cross-section after 2 hours at 1700°C (element analysis by ESD).

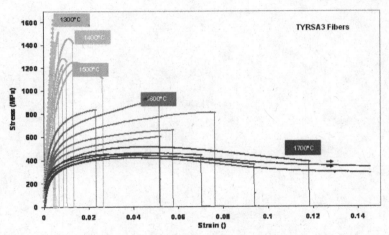

Figure 3 – Stress-Strain curves obtained on TSA3 fibers at high temperature: transition between linear elastic and plastic behaviours.

FOOTNOTES

* Laboratoire des Composites Thermo-Structuraux, CNRS-CEA-SNECMA, Bordeaux, France

REFERENCES

1. Colin C., Gélébart L., Proceedings of 13th ECCM, June 2008, Stockholm, Sweden

2. Sauder C., et al., Composite Science and Technology, 62, pp 499-504, 2002

3. Pailler R., et al., Ann. Chim. Sci. Mat., 30(6), pp 565-578, 2005

4. Sha J.J., et al., Journal of Nuclear Materials, 329-333, pp 592-596, 2004

5. Morimoto T., Composites: Part A, 37, pp 405-412, 2006

6. Sauder C., et al., Carbon, 43, pp 2044-2053, 2005

7. Sauder C., PhD, Université de Bordeaux 1, 2001

8. Scholz R, et al., Journal of Nuclear Materials, 307-311, pp 1098-1101, 2002

9. Sauder C., Référence Interne CEA-LCTS, 2004

10. Dicarlo J.A., et al., Applied Mathematics and Computation 152, pp 473-481, 2003

FRACTURE RESISTANCE OF SILICON CARBIDE COMPOSITES USING VARIOUS NOTCHED SPECIMENS

Takashi Nozawa and Hiroyasu Tanigawa
Japan Atomic Energy Agency
2-4 Shirakata Shirane, Tokai,
Ibaraki 319-1195, Japan

Joon-Soo Park
Institute of Energy Science and Technology, Co., Ltd.
573-3 Okiube, Ube,
Yamaguchi 755-0001, Japan

Akira Kohyama
Institute of Advanced Energy
Kyoto University
Gokasho, Uji, Kyoto 611-0011, Japan

ABSTRACT

One of key issues for the practical application of SiC/SiC composite, a promising candidate of functional and/or structural components for various energy industries, is to develop design basis of quasi-ductile composites. Understanding the crack propagation behavior of this class of composites is, therefore, essential. For this purpose, determining a fracture resistance of composites is undoubtedly important and this paper aims to evaluate this using various notched specimens: the single-edge notched-bend (SENB) specimen and the double notch tensile (DNT) specimen. With the preliminary analysis by the authors, a new methodology to determine the fracture resistance has been developed based on the non-linear fracture mechanics, separately discussing the effect of irreversible energies. By adopting this method to the SENB test results for nano-infiltration transient-eutectic-phase sintered (NITE) SiC/SiC composites, several key findings were obtained: (1) notch insensitivity, (2) no significant size effect, and (3) superior crack tolerance. Similarly, the notch insensitivity was also identified for the DNT test results, depending on the fiber/matrix interfacial bonding strength.

INTRODUCTION

SiC/SiC composites are candidate materials for nuclear energy systems and other various engineering industries due to the inherently good chemical stability at high-temperatures, strength retention, specific strength, low-activation/low after-heat properties. Of many composite types, NITE-SiC/SiC composites and chemical-vapor-infiltration (CVI) SiC/SiC composites are believed to be viable because of excellent baseline mechanical properties [1-3] and proven radiation stability [4]. The good gas tightness of the dense NITE-SiC/SiC composites put an attractive feature to apply it to the gas-cooled systems [5]. Talking about fusion energy applications, several fusion blanket designs like flow channel inserts in the lead-lithium breeder blankets [6, 7] or structural components [8] have been proposed with particular attractiveness.

With the completion of the "proof-of-principle" phase, the R&D on SiC/SiC composites is now shifting to the more pragmatic phase. One of key issues remained is material data-basing. For instance, generation of engineering database deliverable for design activities of a high-temperature operating advanced DEMO reactor is implemented with a primary priority in the R&D on SiC/SiC composites for the Broader Approach (BA) activity [9].

Composites exhibit quasi-ductility, which is quite different from the ductility of metals, since this quasi-ductility occurs as a result of cumulative accumulation of irreversible permanent damages such as interface debonding, fiber pullouts with friction at the fiber/matrix (F/M) interface, and fiber breaks. For the design of composites, the failure behavior, i.e., matrix crack pop-in and crack extension, which gives quasi-ductility in composites, should be evaluated since the functionality and structural stability of composites often deteriorates upon failure initiation. Note that we strictly distinguish failure from fracture; the fracture of composites is referred as an event that the composites break into pieces with no further load sharing. Since this failure behavior depends significantly on the type and distribution of the internal flaws as a crack origin, there would be a significant gap depending on the composite types. A fundamental question is whether there is a meaningful difference among NITE-SiC/SiC composites and other composites like porous CVI-SiC/SiC composites and polymer impregnation and pyrolysis (PIP) SiC/SiC composites. Understanding the crack propagation behavior of composites is, therefore, essential. For this purpose, determining a fracture resistance of composites is undoubtedly important.

This paper aims to evaluate crack propagation behaviors using various types of pre-notched specimens. Specifically, notch sensitivity issue was first evaluated.

Table I. List of SiC/SiC Composites.

ID	NITE-Thick-Coat	NITE-Thin-Coat	PIP-Coat
Typical microstructure			
Fiber	Tyranno-SA3	Tyranno-SA3	Tyranno-SA3
Architecture	UD	UD	S/W
Fiber volume fraction	~0.45	~0.4	~0.3
F/M interface	250nm PyC	None	150nm PyC
Matrix	NITE-SiC	NITE-SiC	PIP-SiC
Density	2.96 g/cm^3	3.19 g/cm^3	2.12 g/cm^3
Porosity	~0.05	~0.02	~0.20

EXPERIMENTAL

Materials

Two types of pilot-grade NITE-SiC/SiC composites with F/M interphase were fabricated by Institute of Energy Science and Technology, Co., Ltd. (Table I). For the former composite, a ~250 nm-thick pyrolytic carbon (PyC) interphase was chemically-vapor-deposited (CVD) on the fiber surface prior matrix densification (hereafter "NITE-Thick-Coat"). In contrast, for the latter case, very thin PyC interphase (<50 nm) was designed. However, the F/M interfacial coating did not successfully formed as shown in micrographs in Table 1. The major parts of fibers were uncoated, although very thin PyC was formed for limited number of fibers. In this study, we designate this composite "NITE-Thin-Coat"hereafter. For both composites, a highly-crystalline and near-stoichiometric Tyranno-SA 3rd-grade SiC fiber was uni-directionally reinforced with a fiber volume fraction of 0.4~0.45. For the pilot-grade NITE-SiC/SiC composites, a secondary phase (white contrast in Table. I), which is reportedly an oxide composed of sintering additives such as Al_2O_3, SiO_2 and Y_2O_3 [10], was localized in the matrix, specifically within intra-bundles. This oxide phase would, however, not significantly impact test results at room- temperature. Details of the NITE process are reported elsewhere [1-3].

In contrast, a satin-woven SiC/SiC composite (hereafter "PIP-Coat") was also fabricated by the modified PIP method (Ube Industries). A Tyranno-SA 3rd-grade SiC fiber was applied. A ~100 nm-thick CVD-SiC and a subsequent ~150 nm-thick PyC coating were formed on the fiber surface. Of particular emphasis is that this modified PIP process gives the PIP-SiC matrix of stoichiometric composition (C/Si≈1). Small pores were distributed in the matrix as shown in Table I, resulting in relatively lower density of ~2.1 g/cm^3. Details of the stoichiometric PIP process were described elsewhere [11].

Single-Edge Notched-Bend Test

Figure 1 shows a photo image of the test setup and drawings of the SENB test specimen. Various size specimens with a different initial notch depth were applied. Note that the width to length ratio was fixed for all specimen types. For comparison, un-notched bend specimens with a size of 20 mm ×4 mm ×1.5 mm were tested. The fiber longitudinal direction is set parallel to the specimen longitudinal direction. Test coupons including an artificial notch were machined from the composite plate by a diamond saw. The specimen surfaces were then polished by the standard metallographic technique with a surface finish of ~1 μm for the crack extension observation. The radius of the notch root was approximately ~150 μm. The SENB tests were conducted at room-temperature using an electromechanical testing machine. Test specimens were loaded using a three-point bend fixture with a support span of 16 mm for SENB-1~3 and 32 mm for SENB-4. The crack opening displacement (COD) was measured by the clip-on gauge. A constant crosshead displacement rate was 0.1 mm/min. The unloading/reloading sequences were applied to evaluate the damage accumulation behavior during the tests. In this paper, test results of the only NITE-Thick-Coat composite are reported. No data is presently available for NITE-Thin-Coat. In contrast, all tests for PIP-Coat were invalid because of the compressive failure at the loading points.

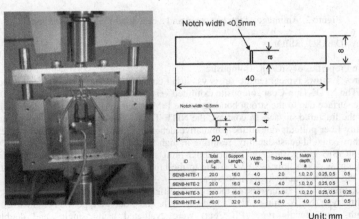

ID	Total Length, L_t	Support Length, L	Width, W	Thickness, t	Notch depth, a	a/W	t/W
SENB-NITE-1	20.0	16.0	4.0	2.0	1.0, 2.0	0.25, 0.5	0.5
SENB-NITE-2	20.0	16.0	4.0	4.0	1.0, 2.0	0.25, 0.5	1
SENB-NITE-3	20.0	16.0	4.0	1.0	1.0, 2.0	0.25, 0.5	0.25
SENB-NITE-4	40.0	32.0	8.0	4.0	4.0	0.5	0.5

Unit: mm

Figure 1. An image of the test setup and drawings of SENB test coupons.

Double Notch Tensile Test

Figure 2 shows a photo image of the test setup and a drawing of the DNT test specimen. Notched tensile specimens with a different initial notch depth were used. For comparison, un-notched tensile specimens with a gauge size of 15 mm ×4 mm ×1.5 mm were tested. The fiber longitudinal direction is set parallel to the specimen longitudinal direction. Test coupons including an artificial notch were machined from the composite plate by a diamond saw. The radius of the notch root was ~150 μm. The DNT tests were conducted at room-temperature using an electromechanical testing machine using a wedge-type gripping device, conventionally used for standard tensile testing. The crack opening displacement was measured by a pair of COD gauges. The average of two readings was applied in calculation. A constant crosshead displacement rate was 0.5 mm/min. The unloading/ reloading sequences were applied to evaluate the damage accumulation behavior during the tests. The DNT tests were applied to all SiC/SiC composite types listed in Table 1.

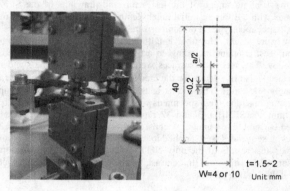

Figure 2. An image of the test setup and a drawing of DNT test coupons.

RESULTS AND DISCUSSION

Fast Fracture Properties of SiC/SiC Composites

Figures 3 shows typical tensile stress vs. strain curves of SiC/SiC composites by the standard tensile test. The NITE-Thin-Coat composite exhibited very brittle fracture. The SEM image apparently shows brittle surface due to the strong bonding at the F/M interface (Figure 4). No fiber pullouts were observed in the fractured surface. In contrast, the NITE-Thick-Coat composite showed quasi-ductility, showing many fiber pullouts due to the cumulative debonding until fracture. Similarly, the PIP-Coat composite showed good quasi-ductility, resulting in high fracture strain (~0.9 %). However, compared with NITE-SiC/SiC composites, the proportional limit stress (PLS), i.e., an equivalent stress when matrix cracking, was very low (~30 MPa) due to the brittle PIP-SiC matrix. For both NITE-SiC/SiC composites, the PLS was quite high (~150 MPa).

Crack propagation behaviors of these three different types of SiC/SiC composites: 1) high-strength, brittle type (NITE-Thin-Coat), 2) high-strength, quasi-ductile type (NITE-Thick-Coat) and 3) poor strength, quasi-ductile type (PIP-Coat) were evaluated using single- and double-notched specimens later. Specifically, the fracture resistance of NITE-Thick-Coat was preliminarily determined by the proposing analytical method based on the non-linear fracture mechanics.

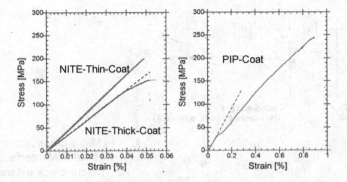

Figure 3. Tensile fracture behaviors of pilot-grade NITE- and PIP-SiC/SiC composites.

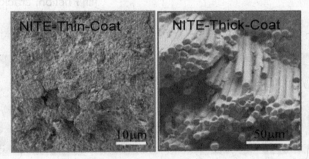

Figure 4. Typical fracture surface images of pilot-grade NITE-SiC/SiC composites.

Single-Edge Notched-Bend Test

Figure 5 shows a typical load vs. crack opening displacement curve of SENB tests. Micrographs of the specimen surfaces in each loading stage were also presented in the figure. Additionally, crack length data measured by the direct reading from the replica films were also plotted. From Fig. 5, apparent three damage stages in fracture behavior were identified:

(1) initial elastic segment followed by the non-linear damage accumulation stage due to micro-cracking not macro-cracking,
(2) macro-cracking from the notch root with a rapid load drop at the maximum load,
(3) load transferring by friction at the fiber/matrix interface, coupled with crack branching and fiber breaks until complete fracture.

In the first stage, many micro-cracks were formed along the fiber longitudinal direction near the root of the notch, however, no macro cracking in the loading direction from the root of notch was observed. In the second stage, macro-crack length increased linearly with increasing crack opening displacement. Compared with the rapid crack extension in the second stage, a mild increase of the crack length was obtained in the third stage, i.e., in the mixed damage process.

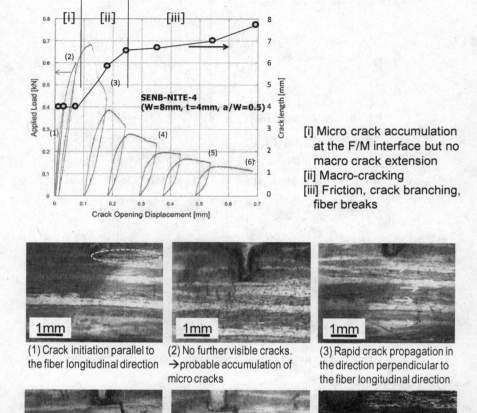

[i] Micro crack accumulation at the F/M interface but no macro crack extension
[ii] Macro-cracking
[iii] Friction, crack branching, fiber breaks

(1) Crack initiation parallel to the fiber longitudinal direction

(2) No further visible cracks. →probable accumulation of micro cracks

(3) Rapid crack propagation in the direction perpendicular to the fiber longitudinal direction

(4) Crack branching and load transfer via interfacial friction

(5) Further branching and fiber sliding at the interface

(6) No breakage into pieces due to huge interfacial contribution

Figure 5. Typical SENB fracture behavior of the NITE-Coat composite.

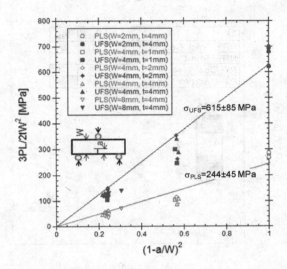

Figure 6. Notch Sensitivity of the NITE-Thick-Coat composite by SENB tests.

From Fig. 5, two characteristic parameters: a proportional limit strength (PLS) as an initiation load of micro-cracking and a maximum flexural strength (UFS) as an initiation load of macro-cracking coupled with cumulative micro-crack formations were identified and they were plotted with respect to a function of the notch depth to width ratio as shown in Figure 6. In the figure, both PLS and UFS were normalized as a flexural stress form. One of key conclusions is that the slopes of the linear fit give unique PLS and UFS regardless of the presence of initial notches. In short, an apparent notch insensitivity of the NITE-Coat composite was clearly demonstrated. Additionally, Fig. 6 also indicates that the notch insensitivity was seemingly independent of specimen size.

Double Notch Tensile Test

Similar to the SENB test results, the micro-cracking load (P_{PLS}) and the macro-cracking load (P_{max}) normalized by the maximum specimen cross-sectional area were plotted as a function of the notch depth to specimen width ratio for all types of SiC/SiC composites (Figure 7). Note that the proportional limit load of the brittle NITE-SiC/SiC composite was equivalent to the maximum load. Of particular emphasis is that both normalized micro-cracking load and macro-cracking load were proportional to $1-a/W$ for NITE-Thick-Coat and PIP-Coat. This means that any SiC/SiC composites with PyC interphase exhibited notch insensitivity regardless of the fabrication process. With this fact, PLS and UTS of composites can be easily estimated by the results of notched specimens. Figure 7 also shows no significant size effect for the PIP-Coat composite. This result was quite similar to that of SENB test results.

In contrast, the limited number of data indicates the notch-sensitivity of the NITE-Thin-Coat composite like brittle ceramics. From Fig. 7, the strength of NITE-Thin-Coat exhibits non-linear relation to $1-a/W$. This is because the strong bonding at the F/M interface does not role as crack branching (Fig. 4).

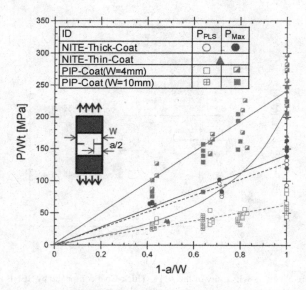

Figure 7. Notch Sensitivity of SiC/SiC composites by DNT tests.

Failure Energy Analysis

The fact of notch insensitivity is very important when considering a design basis of this class of composites since this implies an apparent correlation between stress and fracture energy criterions in fracture of the composites. When this relationship is clearly identified, unique failure criterion can be utilized for the component design. An energy-based criterion is proposed and preliminarily evaluated for NITE-Thick-Coat and PIP-Coat materials hereafter.

Test results apparently show quasi-ductility of composites with F/M interphase, i.e., energy consumption during irreversible damage accumulation beyond matrix cracking. Therefore, considering separately contributions from irreversible energies such as interfacial friction, thermal-residual strain energy, and fiber breaks becomes a key issue. Because of this quasi-ductility, taking an analytical model based on the non-linear fracture mechanics [12, 13] is reasonable and the analytical method has been developed by the authors [14]. In this model, the total work during the notched specimen test (w) is expressed as:

$$w = U_e + U_{fr} + U_r + \Gamma,$$ (1)

where elastic energy (U_e), friction energy at the interface (U_{fr}), residual strain energy (U_r), and crack surface formation energy (Γ) are defined in Figure 8 and they are plotted as a function of crack opening displacement (Fig. 5). Note that the crack surface formation energy includes micro- and macro-crack forming energies together. Then, the fracture resistance (G) can be defined as:

$$G = \frac{\partial \Gamma}{t \partial a} = \frac{1}{t} \frac{\partial \Gamma}{\partial x} \frac{\partial x}{\partial a},$$ (2)

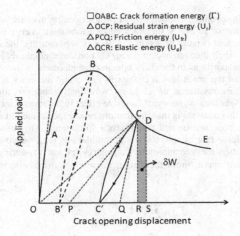

Figure 8. Schematic illustration of load-crack opening displacement curve.

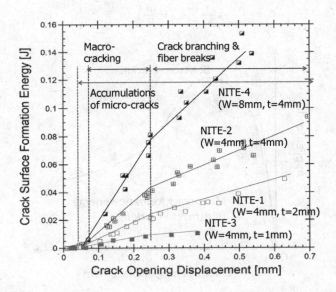

Figure 9. Crack surface formation energy vs. crack opening displacement in SENB tests.

where the crack length (a), specimen thickness (t) and crack opening displacement (x).

Following the definition in Fig. 8, the crack surface formation energy was plotted as a function of crack opening displacement (Figure 9). From Fig. 9, it is apparent that the crack surface formation energy rapidly increased when damage accumulated. Of particular emphasis is that the crack surface formation energy seems proportional to the crack opening displacement in the second stage. Combining the secondary stage data of the crack length change in Fig. 5 and the crack surface formation energy change in Fig. 9, a fracture resistance of ~5 kJ/m² was finally obtained for the NITE-Thick-Coat composite using Eq. (2). Specifically, no significant size effect was obtained for certain test conditions.

One drawback of this analysis is that the fracture resistance defined in Eq. (2) cannot perfectly distinguish contributions from micro- and macro-crack formations. This issue is then discussed by considering an energy release rate by micro-crack accumulations until macro-crack causes. Figure 10 shows a schematic illustration for determination of an energy release rate during micro-crack formation from the load-displacement curve. From the figure, micro-crack surface formation energy (Γ_m) was empirically obtained as:

$$\Gamma_m \cong \sum_{i=1}^{n} C_i \left(1 - \frac{a}{W}\right)^i,$$ (3)

where C_i (i=1,2, ..., n) are constants. Then, an energy release rate (G_c) is then defined as:

$$G_c \equiv -\frac{\Delta \Gamma_m}{t \Delta a} \cong \frac{1}{Wt} \sum_{i=1}^{n} i C_i \left(1 - \frac{a}{W}\right)^{i-1},$$ (4)

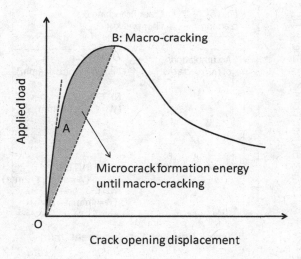

Figure 10. Definition of microcrack formation energy until macro-cracking.

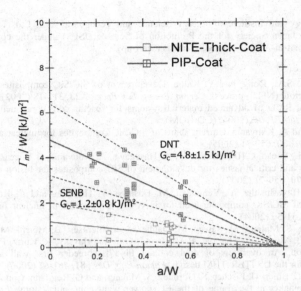

Figure 11. Micro-crack formation energy vs. initial notch depth.

Figure 11 shows crack surface formation energy with respect to the initial notch depth to width ratio using various specimens. Simply applying linear fit in Fig. 11, we obtained an energy release rate of ~1.2 kJ/m^2 for NITE-Thick-Coat regardless of specimen size. This value is quite lower than that of PIP-Coat, ~4.8 kJ/m^2. This implies more cracking resistant for pilot-grade NITE-SiC/SiC composites, since this energy release rate means crack density in damage process. In contrast, it is concluded that PIP-SiC/SiC composites are more damage tolerant. Eventually, the actual fracture resistance for macro-cracking of ~3.8 kJ/m^2 was estimated for NITE-Thick-Coat.

CONCLUSIONS

Silicon carbide composites are promising candidate materials for fusion reactors and other high-temperature applications. This paper provided a present status in development of design basis for quasi-ductile SiC/SiC composites as a structural application. Specifically, the failure behavior, i.e., matrix cracking behavior, of NITE-SiC/SiC composites, as well as PIP-SiC/SiC composites for comparison, was evaluated. With a fact of the notch-insensitivity, a correlation between the stress and energy criterions was implied for both SENB and DNT tests. This notch insensitivity depends significantly on the fiber/matrix interfacial bond strength. Additionally, this study provided a rigorous solution to distinguish contributions of micro-cracking from macro crack extension energy, giving a fracture resistance as an energy failure criterion. Applying this method, good crack resistance of pilot-grade NITE-SiC/SiC composites was demonstrated.

ACKNOWLEDGEMENTS

This research was supported by the framework of the Agreement between The Government of Japan and the European Atomic Energy Community for The Joint Implementation of The Broader Approach Activities in the Field of Fusion Energy Research and collaboration between Japan Atomic

Energy Agency and Institute of Energy Science and Technology Co., Ltd. This work was partly sponsored by Japan Society for the Promotion of Science (JSPS) under the contract of KAKENHI 19860089, Grant-in-Aid for Young Scientists (Start-up).

REFERENCES
[1]A. Kohyama, S.M. Dong, and Y. Katoh, Development of SiC/SiC composites by nano-infiltration transient eutectoid (NITE) process, *Ceram. Engng. Sci. Proc.*, **A23**, 311-18 (2002).
[2]A. Kohyama, R&D of advanced material systems for reactor core component of gas cooled fast reactor, *Proc. ICAPP'05*, (2005) (CD-ROM).
[3]T. Hinoki, and A. Kohyama, Current status of SiC/SiC composites for nuclear applications, *A., Ann. Chim. Sci. Mat.*, **30**, 659-71 (2005).
[4]Y. Katoh, L.L. Snead, C.H. Henager, Jr., A. Hasegawa, A. Kohyama, B. Riccardi, and H. Hegeman, Current status and critical issues for development of SiC composites for fusion applications, *J. Nucl. Mater.*, **367-370**, 659-71 (2007).
[5]T. Hino, E. Hayashishita, Y. Yamauchi, M. Hashiba, Y. Hirohata, and A. Kohyama, Helium gas permeability of SiC/SiC composite used for in-vessel components of nuclear fusion reactor, *Fusion Eng. Des.*, **73**, 51-56 (2004).
[6]C.P.C. Wong, S. Malang, M. Sawan, M. Dagher, S. Smolentsev, B. Merrill, M. Youssef, S. Reyes, D.K. Sze, N.B. Morley, S. Sharafat, P. Calderoni, G. Sviatoslavsky, R. Kurtz, P. Fogarty, S. Zinkle, and M. Abdou, An overview of dual coolant Pb-17Li breeder first wall and blanket concept development for the US ITER-TBM design, *Fusion Eng. Des.*, **81**, 461-67 (2006).
[7]P. Norajitra, L. Bühler, U. Fischer, S. Gordeev, S. Malang, and G. Reimann, Conceptual design of the dual-coolant blanket in the frame of the EU power plant conceptual study, *Fusion Eng. Des.*, **69**, 669-73 (2003).
[8]A.R. Raffray, L. El-Guebaly, S. Malang, I. Sviatoslavsky, M.S. Tillack, X. Wang, and The ARIES Team, Advanced power core system for the ARIES-AT power plant, *Fusion Eng. Des.*, **82**, 217-36 (2007).
[9]T. Nishitani, H. Tanigawa, S. Jitsukawa, T. Nozawa, K. Hayashi, T. Yamanishi, K. Tsuchiya, A. Mslang, N. Baluc, A. Pizzuto, E.R. Hodgson, R. Laesser, M. Gasparotto, A. Kohyama, R. Kasada, T. Shikama, H. Takatsu, and M. Araki, Fusion material development program in the broader approach activities, *J. Nucl. Mater.*, (2008) (in press).
[10]H. Kishimoto, K. Ozawa, O. Hashitomi, and A. Kohyama, Microstructural evolution analysis of NITE SiC/SiC composite using TEM examination and dual-ion irradiation, *J. Nucl. Mater.*, **367-370**, 748-52 (2007).
[11]T. Nozawa, K. Ozawa, Y. Katoh, and A. Kohyama, Heat treatment effect on microstructures and mechanical properties of stoichiometric PIP SiC/SiC composites, *Mater. Trans.*, **45**, 307-10 (2004).
[12]V. Kostopoulos, Y.P. Markopoulos, Y.Z. Pappas, and S.D. Peteves, Fracture energy measurements of 2-D carbon/carbon composites, *J. Eur. Ceram. Soc.*, **18**, 69-79 (1998).
[13]M. Sakai, K. Urashima, and M. Inagaki, Energy principle of elastic-plastic fracture and its application to the fracture mechanics of a polycrystalline graphite, *J. Am. Ceram. Soc.*, **66**, 868-74 (1983).
[14]T. Nozawa, T. Hinoki, A. Kohyama, and H. Tanigawa, Evaluation on Failure Resistance to Develop Design Basis for Quasi-Ductile Silicon Carbide Composites for Fusion Application, *Preprints of the 22nd IAEA Fusion Energy Conference*, FT/P2-17 (2008).

OPTIMIZATION OF AN INTERPHASE THICKNESS IN HOT-PRESSED SiCf/SiC COMPOSITES

Weon-Ju Kim, Jong Hoon Lee, Dang-Hyok Yoon*, Ji Yeon Park
Nuclear Materials Research Division, Korea Atomic Energy Research Institute, 1045 Daedeokdaero, Yuseong-gu, Daejeon 305-353, Republic of Korea
*School of Materials Science and Engineering, Yeungnam University, 214-1 Dae-dong, Gyeongsan, Gyeongbuk 712-749, Republic of Korea

ABSTRACT

The effects of the thickness of PyC interphase on the properties of hot-pressed SiCf/SiC composites were investigated. Slurries of SiC nano-powders containing $Al_2O_3/Y_2O_3/MgO$ sintering additives were infiltrated into the SiC fiber preform with various PyC thicknesses of 200–1000 nm. Densification behavior, microstructural evolution, and mechanical properties of the hot-pressed composites were evaluated. The hot-pressed composite revealed an optimum interphase thickness which was larger than the CVI composite. Density of the hot-pressed composites decreased as the PyC thickness increased. Decrease of the density with the PyC thickness was attributed to the limitation of the slurry infiltration within the SiC fiber bundles. The SiCf/SiC composites with the PyC thicknesses of 200 and 400 nm showed brittle fracture behaviors due to the elimination of the PyC interphase by the reaction between the oxide phases and PyC during fabrication. However, the composites with the PyC coating of more than 600 nm exhibited pseudo-ductile fracture behaviors with higher strengths.

INTRODUCTION

Silicon carbide fiber-reinforced silicon carbide matrix (SiCf/SiC) composites are being considered as structural components for fusion and advanced fission reactors due to their excellent high temperature mechanical properties, chemical stability and low radioactivity under a neutron irradiation [1,2]. Among the various fabrication methods of the SiCf/SiC composites, the chemical vapor infiltration (CVI) process is the most commonly used method which can produce a high quality SiC matrix [3]. The highly crystalline and stoichiometric SiC matrix renders the CVI composite to be highly stable under an irradiation. Therefore, the CVI composite would be the best choice for an application to in-core components under a harsh neutron environment. A hot-pressed composite with a low porosity could also be considered for components, especially those requiring a gas hermeticity, e.g., fuel pins of gas fast reactors [4] and heat exchangers of high-temperature reactors [5]. It has recently been reported that the use of nano-sized SiC powder and Al_2O_3/Y_2O_3 sintering additives forming transient eutectics with SiO_2 (NITE process) resulted in SiCf/SiC composites with a very low residual porosity, high mechanical and thermal properties [6].

The mechanical properties of SiCf/SiC composites largely depend on the thickness of an interphase [7]. The interphase, normally a pyrolytic carbon (PyC), renders a non-catastrophic fracture behavior in the composites. In the case of SiCf/SiC composites fabricated by the CVI method, the PyC thickness of about 200 nm has been known to be an optimum value even though it could be dependent on the type of the SiC fiber used and the detailed process conditions [8–10]. However, in the SiCf/SiC composites fabricated by a slurry infiltration and hot-pressing, the optimum PyC thickness would be different from the CVI composites because of complex reasons such as a possible reaction between the PyC and the oxide phases, e.g., sintering additives and a native silica layer present on the SiC nano-particles, and the application of pressure during fabrication [11]. In this study, we have investigated the effect of the PyC thickness on the properties of hot-pressed SiCf/SiC composites. Slurries of SiC nano-powders containing $Al_2O_3/Y_2O_3/MgO$ sintering additives were infiltrated into the SiC fiber preform with various PyC thicknesses of 200–1000 nm. Densification behavior, microstructural evolution and mechanical properties of these composites were described.

EXPERIMENTAL PROCEDURE

Disk shape fabrics with diameters of 50 mm were punched out from a plain weave Tyranno SA3 fabric (Ube Industries, Japan). The PyC interphase with a thickness of 200–1000 nm was deposited on the fiber surface by a decomposition of methane (CH$_4$) at 1100°C for various times with a deposition pressure of 12 kPa. Nano-sized β-SiC powder (4620KE, Nanostructure & Amorphous Materials Inc., USA) with an average particle size of 52 nm was used in this study. The SiC powder was mixed with sintering additives (Al$_2$O$_3$, Y$_2$O$_3$ and MgO) in a liquid mixture of toluene and ethanol. The average particle sizes of Al$_2$O$_3$, Y$_2$O$_3$ and MgO were 150, 220 and 160 nm, respectively. The amounts of sintering additives were Al$_2$O$_3$ = 6.4 wt%, Y$_2$O$_3$ = 2.6 wt% and MgO = 1.0 wt%. Hypermer-KD and polyvinyl butyral (PVB) were used as a dispersant and a binder, respectively. Each PyC-coated fabric was infiltrated with the slurry in vacuum. After drying and binder burn-out, fifteen layers of the slurry-infiltrated fabrics were stacked with a fabric layer orientation of 0°/90° and restrained in a graphite die. The stacked fabrics were hot-pressed at 1750°C for 2 h in Ar atmosphere under a pressure of 20 MPa.

The hot-pressed composites were machined into 4.0 mm × 2.0 mm × 40 mm flexural bar specimens. Bulk density of the bar specimen was determined by the Archimedes method using distilled water as an immersion medium. Microstructures of the fabricated composites were observed using a scanning electron microscopy (SEM, JS-5200, Jeol, Japan). The flexural strength was measured by the three-point bending test. The span length and the cross-head speed were 30 mm and 0.5 mm/min, respectively.

RESULTS AND DISCUSSION

We varied the deposition time from 150 to 600 min to control the thickness of the PyC interphase. Figure 1 shows the thickness of the PyC layer deposited on the SiC fiber with various deposition times. The PyC layer was coated on the preform stacked with fifteen layers of the SiC fabric and the thickness of the PyC layer was measured along the thickness direction of the preform. The thickness variation of the PyC layer was not significant throughout the preform. A typical high-resolution TEM image (HRTEM) of the PyC interphase deposited at a same condition to the current

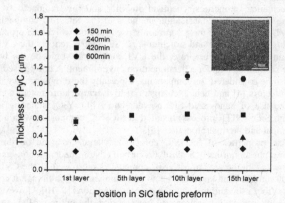

Figure 1. Thickness of PyC interphase along the thickness direction of SiC fabric preform with various deposition times. The inset image represents a typical HRTEM microstructure of the PyC interphase deposited at a similar condition.

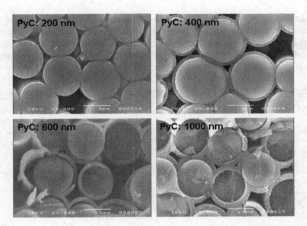

Figure 2. Microstructures of the PyC-coated SiC fibers with various PyC thicknesses.

study is also included in Fig. 1. The HRTEM image shows that the PyC was deposited with a layered structure. Figure 2 shows the cross-sectional SEM microstructures of the PyC-coated SiC fibers with various PyC thicknesses. The SiC fibers within the bundle were coated well and the size of the interfiber space decreased as the deposition time increased.

Each SiC fabric was separated from the PyC-coated preform and vacuum-infiltrated with the SiC slurry containing sintering additives. The slurry-infiltrated fabrics were restacked and hot-pressed after drying and binder burn-out. Figure 3 shows the X-ray diffraction profile of the hot-pressed SiCf/SiC composite with the PyC thickness of 600 nm. The composite consisted of β-SiC phase originated from the matrix and fiber, $Y_3Al_5O_{12}$ (YAG) phase from the sintering additives and a small amount of carbon from the PyC interphase. Typical characteristics of the hot-pressed composites are summarized in Table I. Density of the hot-pressed composites decreased as the PyC thickness increased. The average flexural strength of the composites, however, tended to increase as the PyC thickness increased despite the density decrease. Decrease of the density with the PyC thickness was

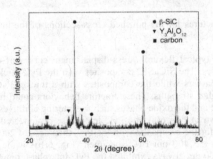

Figure 3. A typical XRD pattern of the hot-pressed SiCf/SiC composite.

attributed to the limitation of the slurry infiltration within the SiC fiber bundles. As shown in Fig. 4, the interbundle matrix regions were nearly fully densified. However, the intrabundle regions contained many pores and the porosity within the fiber bundle increased as the PyC thickness increased. The SiC slurry was not readily infiltrated into the fiber bundle in the thick PyC-coated preform because of the decrease of the interfiber space.

Table I. Characteristics of the Hot-Pressed Composites with Various PyC Thicknesses

Specimen	PyC thickness (nm)	Fiber fraction (vol%)	Bulk density (g/cm^3)	Relative density (%)	Flexural strength (MPa)
PyC200	210–250	52	2.85	90.6	196 ± 18
PyC400	370–380	44	2.69	85.8	223 ± 12
PyC600	570–650	42	2.56	82.9	256 ± 33
PyC1000	930–1100	36	2.51	80.8	282 ± 23

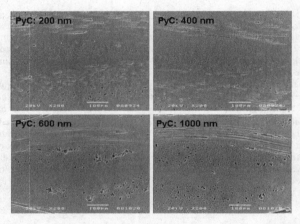

Figure 4. Microstructures for the polished cross-sections of the hot-pressed composites with various PyC thicknesses.

Figure 5 shows the typical flexural stress-displacement curves of the hot-pressed composites with various PyC thicknesses. The SiC$_f$/SiC composites with the PyC thicknesses of 200 and 400 nm showed brittle fracture behaviors while the composites with a thicker PyC exhibited pseudo-ductile fracture behaviors with higher strengths. These fracture behaviors could be attributed to the reaction between the PyC interphase and the oxide phases such as sintering additives and silica contained in the SiC nano-powder. As shown in Fig. 6(a), the PyC interphase disappeared at the region between the fiber and the matrix phase for the composite with the PyC thickness of 200 nm. Some remaining PyC was occasionally found in the 400 nm PyC specimen (Fig. 6(b)) while the composites with PyC thicknesses of more than 600 nm always exhibited the PyC interphase remained at the region between the fiber and the matrix (Fig. 6(c)). The remaining PyC interphase would lead to crack deflections and

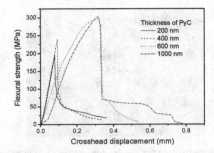

Figure 5. Typical flexural stress-displacement curves of the hot-pressed composites with various PyC thicknesses.

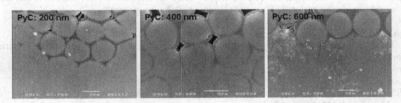

Figure 6. Microstructures for the regions between the fiber bundle and matrix phase of the hot-pressed composites with various PyC thicknesses.

a fiber pull-out, and therefore, a pseudo-ductile fracture behavior. Figure 7 shows the typical fracture surface of the hot-pressed composite with the PyC thickness of 1000 nm, in which the pull-out of fibers is clearly shown. From the viewpoint of the mechanical properties, a PyC coating of more than 600 nm is required to induce a non-brittle fracture behavior. However, there would be a trade-off in the interphase thickness because thick PyC interphases hinder the slurry infiltration into the intrabundle regions. A more efficient way of a slurry infiltration, e.g., an electrophoretic deposition (EPD) method, would be required to increase the density of the intrabundle regions. Our recent results indicate that the EPD method is promising for increasing the composite density. A thin SiC protective coating could also be considered to minimize the interaction between the oxide phases and the PyC interphase during hot-pressing.

CONCLUSIONS

The effects of the PyC thickness on the densification behavior, microstructural evolution and mechanical properties of hot-pressed SiCf/SiC composites were investigated. Slurries of SiC nano-powders containing Al_2O_3/Y_2O_3/MgO sintering additives were infiltrated into the SiC fiber preform with various PyC thicknesses of 200–1000 nm. Density of the hot-pressed composites decreased as the PyC thickness increased. The flexural strength of the composites, however, tended to increase as the PyC thickness increased despite the density decrease. Decrease of the density with the PyC thickness was attributed to the limitation of the slurry infiltration within the SiC fiber bundles. The SiCf/SiC composites with the PyC thicknesses of 200 and 400 nm showed brittle fracture behaviors due to the

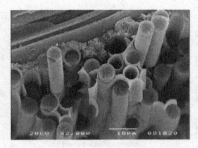

Figure 7. A typical microstructure for the fracture surface of the hot-pressed composite with the PyC thickness of 1000 nm.

elimination of the PyC interphase during fabrication while the composites with more than 600 nm PyC thickness exhibited pseudo-ductile fracture behaviors with higher strengths.

ACKNOWLEDGMENTS

This study was supported by The Ministry of Knowledge Economy through a Materials & Components Technology R&D Program.

REFERENCES

[1]L.L. Snead, R.H. Jones, A. Kohyama, and P. Fenici, Status of Silicon Carbide Composites for Fusion, *J. Nucl. Mater.*, **233-237**, 26-36 (1996).
[2]J.-P. Bonal, A. Kohyama, J. van der Laan, and L.L. Snead, Graphite, Ceramics, and Ceramic Composites for High-Temperature Nuclear Power, *Mater. Res. Soc. Bull.*, **34**, 28-34 (2009).
[3]R. Naslain, Design, Preparation and Properties of Non-Oxide CMCs for Application in Engines and Nuclear Reactors: an Overview, *Comp. Sci. Technol.*, **64**, 155-70 (2004).
[4]A. Kohyama, T. Hinoki, T. Mizuno, J.S. Park, and H. Kishimoto, Advanced GFR Utilizing NITE-SiC/SiC Shield Fuel Pin, Proceedings of ICAPP '06, 2149-56 (2006).
[5]W. Hoffelner, High Temperature Materials - The Challenge for Future Advanced Gas Cooled Reactors, Proceedings of ICAPP '04, 380-7 (2004).
[6]S. Dong, Y. Katoh, and A. Kohyama, Preparation of SiC/SiC Composites by Hot Pressing, Using Tyranno-SA Fiber as Reinforcement, *J. Am. Ceram. Soc.*, **86**, 26-32 (2003).
[7]R. Naslain, The Design of the Fiber-Matrix Interfacial Zone in Ceramic Matrix Composites, *Comp. Part A*, **29A**, 1145-55 (1998).
[8]W. Yang, T. Noda, H. Araki, J. Yu, and A. Kohyama, Mechanical Properties of Several Advanced Tyranno-SA Fiber-Reinforced CVI-SiC Matrix Composites, *Mater. Sci. Eng. A*, **A345**, 28-35 (2003).
[9]T. Hinoki, E. Lara-Curzio, and L.L. Snead, Mechanical Properties of High Purity SiC Fiber-Reinforced CVI-SiC Matrix Composites, *Fus. Sci. Technol.*, **44**, 211-8 (2003).
[10]W.-J. Kim, S.M. Kang, K.H. Park, A. Kohyama, W.-S. Ryu, and J.Y. Park, Fabrication and Ion Irradiation Characteristics of SiC-Based Ceramics for Advanced Nuclear Energy Systems, *J. Kor. Ceram. Soc.*, **42**, 575-81 (2005).
[11]K. Shimoda, J.-S. Park, T. Hinoki, and A. Kohyama, Influence of Pyrolytic Carbon Interface Thickness on Microstructure and Mechanical Properties of SiC/SiC Composites by NITE Process, *Comp. Sci. Technol.*, **68**, 98-105 (2008).

VALIDATION OF RING-ON-RING FLEXURAL TEST FOR NUCLEAR CERAMICS USING MINIATURIZED SPECIMENS

S. Kondo, Y. Katoh, J.W. Kim, L.L. Snead
Materials Science and Technology Division, Oak Ridge National Laboratory
Oak Ridge, Tennessee 37831, USA

ABSTRACT

For the purpose of evaluating fracture strength of the inner pyrocarbon (IPyC) layer of TRISO fuel particles, a miniature disc equibiaxial flexural test technique was developed. It was applied to nuclear graphite specimens in a variety of combination of sample thickness and loading ring diameter to investigate the influences on the stress uniformity within the loading ring. Although similar and relatively high Weibull modulus were observed for all conditions (m ≈ 20), the significant stress concentration associated with the large deflection was observed at the loading location for thinner specimens ($t<0.2$mm). However, the true local fracture stress was successfully estimated from the measured fracture load using finite element analysis. The method for estimating true local fracture stress appear reasonable for evaluating the fracture strength of dense poly-crystalline graphite, and can be used for determination of the statistical parameters for fracture stress of graphite and pyrocarbon.

INTRODUCTION

The failure criteria of the inner pyrocarbon (IPyC) layer are among the critical parameters determining the performance of TRISO fuel particles which is one of the advanced fuel concepts for the very high temperature gas-cooled test reactor (VHTR) [1]. However, a sufficiently reliable and relevant method of strength evaluation for IPyC has not been available to date. Considering that the anticipated primary failure mode of IPyC is equibiaxial tension (initiating from the inner surface) due to internal pressure developed by accumulation of gaseous fission products, equibiaxial flexural testing is perhaps the available technique capable of 1) testing a small specimen that is produced in the fluidized bed chemical vapor deposition (CVD) furnace as used for fuel coating, 2) applying a surface equibiaxial tensile loading onto a sample, 3) testing in a simple and robust manner, making post-irradiation evaluation fairly easy, and 4) producing statistically significant data from a small volume neutron irradiation experiment. Furthermore, the measurement of the strength of brittle materials under bi-axial flexure conditions rather than uniaxial flexure is often considered more reliable, because the maximum tensile stresses occur within the central loading area and spurious edge failures are eliminated. The objective of this work is to develop such an equibiaxial disc flexural test for strength determination of PyC material relevant with IPyC in TRISO fuel particles. In a previous report by the authors [2], it was demonstrated that such a test technique developed is effective to determine the apparent flexural strength of thin specimen of PyC. However, the correction factor for estimating true local stress at the location of fracture origin, where the stress concentration is anticipated, from the apparent stress may be essential. The present document reports the influences of the sample thickness and loading ring diameter on the flexural strength using dummy graphite samples in an attempt to estimate the true local fracture stress at the location of fracture origin.

EXPERIMENTAL PROCEDURE

The equibiaxial flexural test in a "ring-on-ring" configuration, where a disc specimen of ceramic is placed on a support ring and loaded with a smaller diameter coaxial loading ring, was adopted as a method of strength evaluation at ambient temperature. This test method is often utilized for equibiaxial strength of brittle ceramics and the test procedure is standardized in ASTM C1499-05. [3]

A typical disc equibiaxial flexural test utilizes ceramic specimens of diameter of the order of one inch. However, because the evaluation of PyC samples requires testing using much smaller dimensions, a dedicated fixture was designed and fabricated. The section view of the test setup is shown in Fig. 1 and in a set of photographs in Fig. 2. The physical parameters of the test configuration and the condition of testing are summarized in Table. I.

Dummy specimens used were near-isotropic poly-crystalline nuclear graphite (AXF-5Q produced by Poco-Graphite Inc., TX, USA), which has similar Young's modulus to fracture stress ratio to that of PyC. Dimensions of the disc specimens were 6.0 mm (diameter) ×0.10, 0.15, and 0.20 mm (thickness). Whereas, only the tension side was polished by 6 mm diamond plate for most DL251T150 (see the notation in Table I) using Minimet 1000 (Buehler Ltd., IL, USA), both surfaces of discs for DL116T100, DL116T150, DL116T350, DL251T100, L251T200, and L251T350 were polished in the same manner.

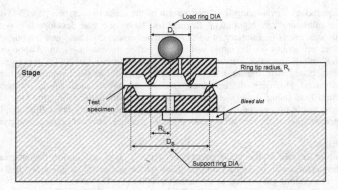

Fig. 1. Section view of the equibiaxial flexural test configuration.

Fig. 2. Photographs of test specimen and fixture before (top-left) and after (bottom-right) testing.

Table I. Physical parameters and condition of the tests.

Sample IDs	Loading Ring DIA. D_L[mm]	Avg. Sample Thickness t[mm]	Std. Dev. for t [mm]	Number of Specimens Tested [pcs.]
DL116T100	1.16	0.101	0.003	28
DL116T150	1.16	0.150	0.002	28
DL116T350	1.16	0.350	0.002	30
DL251T100	2.51	0.101	0.003	28
DL251T150	2.51	0.152	0.004	29
DL251T200	2.51	0.200	0.003	28
DL251T350	2.51	0.349	0.003	30

RESULTS AND DISCUSSION

Estimation of True Local Stress of Thin Disc Samples

Results of the ring-on-ring tests for graphite thin discs (t≤0.2 mm) are summarized in Table II. The load-displacement curves show large deflection (see displacement to thickness ratios in Table II; d/t = 1.3-2.5) and non-linear load-displacement relationship as presented in Fig. 3. Such deviation from linear elastic theory have been previously observed in ring-on-ring testing for glass plate [4], and are associated with the reinforcing effect of outer overhang ($r > D_s$) which can not move freely toward the center. The displacements seem to depend on the loading ring diameter rather than the sample thickness. The failure patterns clearly indicate the stress magnification at the loading location as shown in Fig. 4. In the case of using the loading ring of smaller diameter (D_L = 1.16 mm), some fracture origins are likely located in the area just inside the loading ring. In order to quantitatively evaluate the location of fracture origin, the ratios of r_f to R_L were evaluated in this work, where the r_f is the radial distance between the likely crack origin and the disc center, the R_L is the loading ring radius. The mean r_f to R_L ratios were estimated to be 0.90 for t = 0.1 mm discs and 0.88 for t = 0.15 mm discs, indicating the fracture origin just inside the inner ring. For the tests with larger load ring (D_L = 2.51 mm), the r_f/R_L ratios show higher values (r_f/R_L = 0.92-0.95) compared to smaller load ring case, indicating much significant localized stress near the loading location.

Table II. Summary of results of ring-on-ring test for thin graphite discs.

Sample IDs	Avg. Displacement d[mm]	Avg. Fracture Load F[N]	Mean Fracture Origin r_f/R_L	Weibull Mean Local Stress[1] $\sigma_{p,max}$[MPa]	Std. Dev. for $\sigma_{p,max}$ [MPa]	Weibull Modulus m
DL116T100	0.252	3.13	0.90	134	9.16	18.1
DL116T150	0.259	5.23	0.88	143	9.42	18.8
DL251T100	0.239	5.62	0.95	127	8.61	18.3
DL251T150	0.229	8.79	0.92	126	10.4	14.8
DL251T200	0.226	13.6	0.94	133	8.49	19.5

1) Local flexural stresses were estimated from fracture load by finite element method.

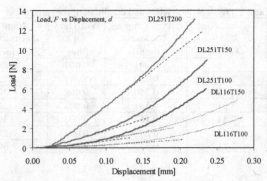

Fig. 3. Typical load-displacement curve for thin disc samples.

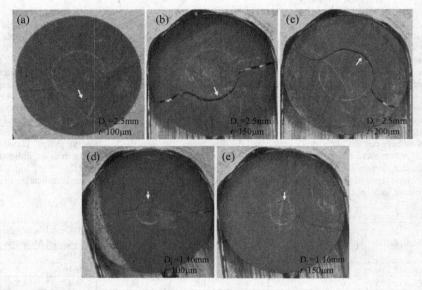

Fig. 4. Failure patterns of graphite discs tested by loading ring with D_L=2.51 for t=;(a)100, (b)150, (c)200μm, or D_L=1.16mm for t=; (d)100, (e)150μm. Likely fracture origins are indicated by arrows (viewed from compression sides for clear vision of loading ring trace).

Kao et al. reported that stress magnification at the loading point is significant when the plate deflection exceeds one-half the specimen thickness [5]. In that case, the equibiaxial stress calculation following ASTM 1499-05 may not be utilized to relate the flexural strength to the flexural load. The local stress at the loading location in a ring-on-ring configuration has been empirically and analytically approximated considering the ratio of deflection to specimen thickness in an attempt to solve the nonlinear plate equations [5, 6]. However, these valuable solutions have no application to thin graphite samples because of the significant deviation from Hooke's law or elastic-plastic fracture behavior. Therefore, finite element analysis was employed in this study to estimate the true local stress from the applied load. Figure 5 shows the radial stress distribution in the ring-on-ring loading for the same material and experimental conditions as Table 1. The stress magnifications beneath (or near) the loading point are computationally well demonstrated by the finite element analysis. The local stress at the loading point is estimated to be twice as large as that at disc center for the t= 100 μm, D_L= 2.51 mm case. For smaller loading ring case (t= 100 μm, D_L= 1.16 mm), somewhat more uniform stress distribution is observed, where the maximum stress is about 1.3 times larger than the center stress. Therefore, one can conclude that the fracture load is governed primarily by the local stress associated with the significant curvature at the loading point in all cases tested.

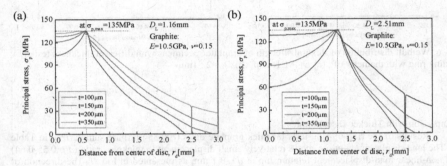

Fig. 5. Radial stress distribution at an approximate flexure stress at the loading location (~135MPa) analyzed by finite element method.

Figure 6 shows the Weibull statistical plots of the flexural strength of graphite discs, where the true local stress at the loading point was estimated from the fracture load using finite element calculation. The Weibull mean strengths are summarized in table II. Although strength were not widely scattered for each case, slight higher strength for using D_L=1.16 mm case (σ_f=134, 143 MPa) than the D_L=2.51 mm case (σ_f=127-133 MPa) was observed. All the measured fracture strength are higher than the commercially reported flexural strength of 86 MPa (standard 4-point bend test [7]) and the typical reported value of 113 MPa for small size bend specimens (2.3 × 6 × 30 mm [8]). For most ceramics, strength depends on the effective stressed area or volume because of the statistical distribution of strength-controlling flaws such as machining flaws [9]. The limited effective-stressed area might modify the results of the present work, though no consistent relationship between bi-axial and uni-axial flexure strength data has been established [10, 11]. It is worth noting that the similar and relatively high Weibull modulus (m= 18.1-19.5) for each test set except for DL251T150 (m= 14.8) were observed. The slight smaller Weibull modulus for DL251T150 is possibly due to the variety of the roughness for the as machined compressive surface and/or the presence of the residual stress at there.

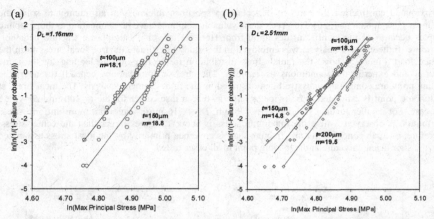

Fig. 6. Weibull distribution of flexural strength of graphite specimens with different thickness tested by loading ring with diameter of ; (a) D_L =1.16mm, (b) D_L =2.51mm.

Comparison with Thicker Disc Samples

Results of the ring-on-ring tests for thick graphite discs (t=0.35 mm) are summarized in Table III. The load-displacement curve shows relatively small displacement to thickness ratio (d/t=0.55-0.64) and non-linear load-displacement relationship at $d>\sim0.1$ mm as presented in Fig. 7. The direction of deviation from the linear load-displacement is opposite of the thinner disc cases. It is well known that strain-stress curve of graphite both in uni-axis static tensile and compressive condition show a non-linear curve due to the decrease in Young's modulus [12]. Therefore, the reinforcing effect of the overhang is insignificant to such an extent that the deviation hide behind the intrinsic degradation of Young's modulus for the cases of thick samples. Figure 8 shows the typical failure patterns of the thicker discs for using the load ring with D_L =1.16 mm and D_L =2.51 mm, respectively. Although, the likely fracture origins, which are indicated by arrows in Fig. 8, are located within inner ring contact lines for both loading ring cases, the flaw characteristics are significantly different. The crack was frequently deflected along the inner side of the load ring for D_L =2.51 mm case, whereas the approximately-straight cracks are seen in the D_L =1.16 mm case. As listed in table III, the mean r_f/R_L ratio for the case of using larger load-ring is three times larger than the smaller load-ring case. The much larger r_f/R_L ratio of 0.6 for DL251T350 may be attributed to the stress magnification that is still remaining inside the inner ring. Because the indirect crack growth normally increases the fracture energy, relatively higher strength might be observed for the D_L =2.51 mm case. Indeed, the strength of DL251T350 was clearly higher than that of the others in spite of the anticipated larger effective stressed area. The very limited reinforcing effects and near uniform stress distributions in a smaller loading ring configuration also observed in our FE analysis for thicker samples. From the both experimental and FE analysis results, the equibiaxial strength is confidently determined to be 129 ± 7.8 MPa for graphite with the identical surface condition.

Table III Summary of results of ring-on-ring test for thick graphite discs.

Sample	Avg. Displacement d[mm]	Avg. Flexure Load F[N]	Mean Fracture Origin r_f/R_L	Weibull Mean Flexural Stress σ_f [MPa]	Std. Dev. for σ_f [MPa]	Weibull Modulus m
DL116T350	0.191	16.8	0.22	129	7.80	20.7
DL251T350	0.224	35.5	0.76	143	6.60	27.3

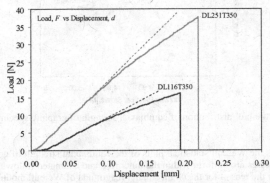

Fig. 7 Typical load-displacement curve for thin disc samples.

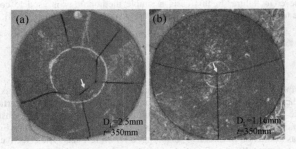

Fig. 8. Failure patterns of graphite discs tested by loading ring in the condition of; (a) D_L=2.51 mm t=350 μm, (b) D_L=1.16mm, 350μm. Likely fracture origins are indicated by arrows (viewed from compression sides for clear vision of loading ring trace).

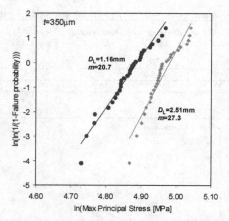

Fig. 9. Weibull distribution of equibiaxial strength of graphite specimens.

Figure 9 shows the Weibull statistical plots of the equibiaxial strength of graphite discs, where assuming principal stresses are uniform within the inner ring and estimated following eq. (7) in ASTM 1499-05 [2]. Although the reason for the slight smaller magnitude of Weibull modulus for DL116T350 (m = 20.7) is not clear, it is similar to the average m value for the test series of thinner discs. This is indicative that a single flaw type is responsible for the failure primarily due to the successful elimination of the edge failure. In this case, a simple methodology is used to convert the strength obtained with one test geometry to strength representing another test geometry: [13],

$$\frac{\sigma_1}{\sigma_2} = \left(\frac{S_1^{\text{eff}}}{S_2^{\text{eff}}}\right)^{-\frac{1}{m}} \qquad (1)$$

,where the subscripts 1 and 2 denote two deferent geometries of test specimens or configuration, S^{eff} is the effective surface area, m is the surface flaw controlling Weibull modulus. Particularly with small size specimen, the effect of stressed surface or volume on failure probability must be considered in order to extrapolate the measured strength to the strength of full size components. Figure 10 shows the flexural strength as a function of effective surface area for all test sets. The effective surface area were estimated based on the results of r_f/R_L ratio evaluation, where the uniform stress were assumed to be ranging from the loading point to lower limit of the standard deviation of r_f/R_L ratio. The slope of the linear-regression fitted line (loading length regressed on strength in this instance) is -1/m, which leads to an additional estimate of the Weibull modulus as 23.1. The additional Weibull modulus is slight larger than the m values obtained from results of each test sets, but it is still comparable to those results, which implies that the total area of stress magnified region may modify the apparent strength due to the size effect. Furthermore, the successfully scaled m value implies that Weibull material parameters such as characteristic strength and scale parameter can be certainly estimated from the ring-on-ring test results for high E/σ materials like PyC.

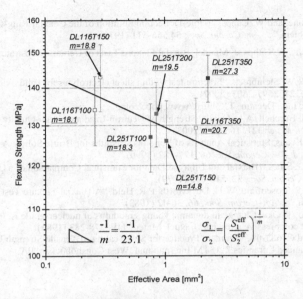

Fig. 10 Flexural strength versus effective surface area for specimens with different thickness.

CONCLUSION

For the purpose of evaluating fracture strength of the inner pyrocarbon (IPyC) layer of TRISO fuel particles, a miniature disc equibiaxial flexural test technique was developed and applied to graphite specimens of varied thickness. It was demonstrated that the test technique is effective to determine the fracture strength of thin specimens of graphite. Moreover, true local fracture stress was estimated from the measured fracture load using results obtained from finite element analysis. The method developed to estimate true local fracture stress values appear reasonable for evaluating the flexural strength of dense poly-crystalline graphite, and can be used for determination of the statistical parameters for fracture stress of graphite and pyrocarbon.

ACKNOWLEDGEMENT

This work was carried out for the Office of Fusion Energy Sciences, US Department of Energy under contract DE-C05-00OR22725 with UT-Battelle, LLC.

REFERENCES

[1] CEGA Report. CEGA-002820, Rev 1 "NP-MHTGR Material Models of Pyrocarbon and Pyrolytic Silicon Carbide," July 1993.
[2] ORNL/TM-2008/164
[3] ASTM Standard C1499-05, "Standard Test Method for Monotonic Equibiaxial Flexural Strength of Advanced Ceramics at Ambient Temperature," ASTM International, West Conshohocken, PA.

[4] R. Kao, N. Perrone, and W. Capps, "Large-Deflection Solution of the Coaxial-Ring-Circular-Glass-Plate Flexure Problem," *J. Am. Ceram. Soc.*, **54**,566-571 (1971).

[5] J.B. Wacgtman Jr., W. Capps, J. Mandel, Biaxial Flexure Tests of Ceramic Substrates, *J. Mater*. **7**, 188-194 (1972).

[6] J. Malzbender, R.W. Steinbrech, "Fracture test of thin sheet electrolytes for solid oxide fuel cells," *J. Euro. Ceram. Soc.* **27**, 2597-2603 (2007).

[7] POCO Graphite, Inc., Decatur, TX. http://www.poco.com/

[8] L.L. Snead, T.D. Burchell, A.L. Qualls, Strength of Neutron-Irradiated High-Quality 3D carbon fiber composite, *J. Nucl. Mater* **321**, 165-169 (2003).

[9] J. Lamon, A.G. Evans, Statistical Analysis of Bending Strengths for Brittle Solids: A Multiaxial Fracture Problem, *J. Am. Ceram. Soc.* **66**, 177-182 (1990).

[10] M.N. Giovan, G. Sines, "Biaxial and Uniaxial Data for Statistical Comparison of a Ceramics's Strength," *J. Am. Ceram. Soc.*, **52**, 510-515 (1979).

[11] D.K. Shetty, A.R. Rosenfield, W.H. Duckworth, P.R. Held, "A Biaxial-Flexure Test for Evaluating Ceramics Strength," *J. Am. Ceram. Soc.*, **66**, 36-42 (1983).

[12] S. Yoda, M. Eto, T. Oku, Change in dynamic young's modulus of nuclear-grade isotropic graphite during tensile and compressive stressing, J. Nucl. Mater., 119, 278-283 (1983).

[13] ASTM Standard C1683-08, "Standard Practice for Size Scaling of Tensile Strength Using Weibull Statistics for Advanced Ceramics," ASTM International, West Conshohocken, PA.

Material and
Component Processing

DESIGN, FABRICATION, AND TESTING OF SILICON INFILTRATED CERAMIC PLATE-TYPE HEAT EXCHANGERS

J. Schmidt, M. Scheiffele, M. Crippa, German Aerospace Center (DLR), Stuttgart, Germany
P. F. Peterson, University of California, Berkeley, USA
K. Sridharan, Y. Chen, L.C. Olson, M.H. Anderson and T.R. Allen, University of Wisconsin, USA

ABSTRACT

A novel concept for hydrogen production has been reported by the US Department of Energy, which combines the use of heat from a nuclear power plant (a Generation IV reactor) or a solar power tower for the production of hydrogen in a thermo-chemical reaction. Ceramic heat exchangers (HX) provide a promising technology for this concept. Novel plate-type HXs with high power densities are proposed, which are based on novel integrated flow-channel designs. The main purpose of this study is the investigation of net-shape fabrication to prototypical HX components based on these designs. To achieve net-shape plates, dry powder mixtures were molded by axial pressing. The joining to the prototypical 3D HX stack was accomplished by lamination followed by pyrolysis at temperatures of up to 1650 ℃. Due to the use of carbon fibers the shrinkage could be controlled and reduced to about 5 %. Finally, accurate silicon melt infiltration by using the wick method into the porous C/C preforms led to dense C/SiSiC ceramics. Microstructural investigations and flexural strength measurements were performed to demonstrate the homogeneity of the ceramic and the quality of the joinings. The gas-tightness of the ceramic composites to helium has been qualified by gas-leakage tests. Corrosion tests with C/SiSiC coupons, both with and without a CVD pyrocarbon-SiC protective coating (bilayer) were performed using a ternary eutectic fluoride salt of LiF, NaF, and KF (FLiNaK) as the intermediate heat transfer fluid. While SiC is vulnerable to corrosion by the salt, such a coating offers a high degree of protection to the ceramic substrate.

INTRODUCTION

Due to the increasing energy demand there is strong need for renewable energy opportunities as an alternative to fossil fuel. The US governmental programs support the development of high temperature heat exchangers (HTHX) for hydrogen production and electrical energy conversion. The projects also support the development of Generation IV (Gen IV) reactors, since nuclear energy is emission free and can provide the high temperatures required for water dissociation technologies. Solar power towers provide an alternative technology to collect high temperature heat. An overall objective is to reduce fossil fuel consumption by supplying non-fossil hydrogen to oil refineries, and ultimately to replace fossil fuel through hydrogen, e.g. for hydrogen-fuel vehicles, by using efficient production technologies. There are two approaches for the production of hydrogen using high-temperature heat: thermally

95

assisted electrolysis and thermo chemical processes, e.g. the sulphur-iodine (S-I) cycle. Both methods could have significantly higher efficiencies than could be achieved with electricity generation and conventional electrolysis.

Very High Temperature Reactors (VHTR's) have high priority among the U.S. reactor concepts of Next Generation Nuclear Power Plants (NGNP), since they can operate at very high temperatures (above 850 ℃) producing 600-2400 megawatts of thermal power [1]. The materials used in the VHTR must withstand very high temperature, intense neutron radiation and corrosive environments. For high-temperature intermediate heat transport, liquid salts are a desirable heat transfer fluid due to their high volumetric heat capacity. For utilizing high temperature process heat from a nuclear reactor for hydrogen production there will be a need for a long pipe of 100 meters or more for the transport of this liquid-salt coolant from the nuclear power plant to the intermediate heat exchanger (IHX) or thermo-chemical plant. For this purpose efficient heat transfer fluids are required. Besides helium, molten fluoride and chloride salt are being considered as heat transport fluids [2]. The material screening focuses on salt compositions with high chemical stability for T > 800 ℃, melting points T < 525 ℃, low vapor pressure, and compatibility with alloys, graphite and ceramics as needed for the heat transfer loop. Impurities, temperature gradients and activity gradients might increase liquid salt corrosion problems. In addition to the LiF/NaF/KF salt (FLiNaK), future high temperature corrosion tests will also study LiCl-KCl-MgCl$_2$ salts, since they also have the potential to meet these basic requirements and are relatively inexpensive [3].

The plate-type HX with small flow channels provides a good approach for constructing an IHX because high power density can be obtained. The off-set fin (OSF) plate design is already proposed, which should enable a very large heat transfer area density and an effective counter current flow. The complete HX module is based on alternating liquid salt and helium plates, which are joined together. The channels should not exceed 3 mm in height, 10 mm in length, and 3 mm in width. The schematic plate design with integrated flow channels is shown in Figure 1. The total HX dimension of such a 3D block should not exceed 0.9 x 1 x 1 m^3. Due to the compact design low stresses were found through FEM simulations, so that safety factors of up to 8 could be achieved [4]. This design is also of considerable interest for HX, which can be used under corrosive environments of externally fired combined cycle (EFCC) processes in power plants [5].

As alternatives to the OSF concept, different shell HX, plate HX, and decomposer designs with varying complexities are also under investigation. All these designs can be processed with ceramic powders by using the Laminated Object Manufacturing (LOM) technique. Single channel, hexagonal, and diamond-like channel designs have been proposed with a channel offset of about 50-100 % [5-7].

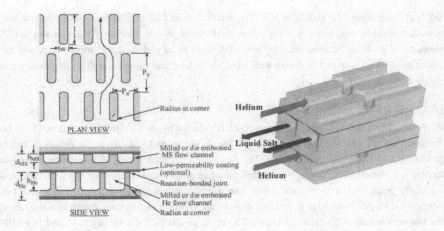

Figure 1. Schematic off-set fin (OSF) plate design enabling a very large heat transfer area density and an effective counter current flow for different media [4].

HX are the key energy conversion components for thermally driven hydrogen production. The success of the Gen IV project relies critically on the design and performance of these system components. Promising material candidates cover high temperature nickel-based alloys and ferritic steels as well as advanced carbon and SiC based composites. The primary coolant of the NGNP reactor will be helium. Advanced reactors, after the NGNP, may also be designed to use liquid salt as the primary coolant. The IHX would then preferably consist of helium-to-liquid-salt or helium-to-helium modules housed within a gas-tight vessel and would operate at temperatures in the range of 600 to 1000 $°C$. For the helium to liquid-salt configuration, the IHX would operate with a pressure difference from 6 to 8 MPa, while for the helium-to-helium configuration the IHX would operate in pressure balance and the process heat exchangers must accommodate the pressure differential. Another critical area of the S-I cycle is the point of interaction between the heat transfer medium and the sulphuric acid decomposition section. An IHX design is required, which optimizes thermal efficiency while being capable of withstanding the corrosive environment.

If such a HX is primarily used as a decomposer for corrosive sulphuric acid, it should maintain appropriate conditions for the chemical dissociation reactions at T > 850 $°C$. The materials must be gas-tight, corrosion resistant and should exhibit high fracture and creep strength and must be capable of operating in the temperature range of 800-1000 $°C$. The material must almost maintain full mechanical strength in this temperature range. The materials must be also cheap and preferably fabricated in net-shape design. Coatings must be applied to achieve gas hermeticity and corrosion resistance against the

S-I feed components like H_2SO_4 or SO_3. The material should be also thermo-shock resistant due to thermal transients, which might occur when the flow of process fluid or coolant is stopped. The temperature gradients between the inlet and outlet could be as high as 400 ℃ so that there will be a strong need for a compact HX design with a low stress level and ceramics with an improved fracture toughness [4].

HEAT EXCHANGER TEST DESIGN

Prior to the accomplishment of the proposed HX design by the UC Berkley, a test mold has been designed and manufactured by DLR to check the feasibility and viability of the net shape molding process, which is the first process step. The fins should be molded with high accuracy and without having any defects. The design of the steel mold (figure 2) covers different types of fins, which especially differ in their length and width. The widths were chosen to be 2 mm, 3 mm, and 4 mm, whereas the lengths were 10 mm, 20 mm, 30 mm, and 50 mm. The following length to width ratios of the fins were adjusted: 5 (10/2), 3.3 (10/3), 2.5 (10/4), 10 (20/2), 5 (20/4), 10 (30/3), and 16.6 (50/3). The depth of all the integrated channels was fixed to 4 mm. A chamfer (1x45°) and a draft angle (3°) were machined into each cavity of the mold to enable the simple deforming of the Wood Based Composite (WBC) plate after the axial pressing process.

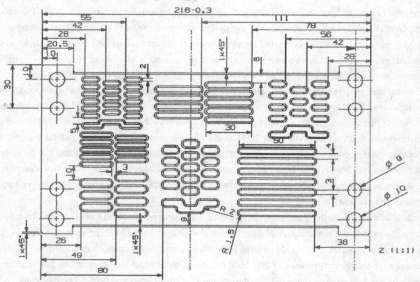

Figure 2. Technical drawing of the test design with different fins and channel sizes to perform the steel mold (dim. 216 x 115 x 4 mm³) needed for the axial net shape pressing.

CONCEPT FOR THE CERAMIC HEAT EXCHANGER FABRICATION

Powder compounding and net shape molding process

For the molding process a compound was used, which consisted of wood powder (Ø < 50 μm), phenolic resin (Ø < 50 μm) and pitch based carbon fibres (l < 370 μm) as the filler. All components were mixed homogeneously in the dry state. The mass ratio of the components was adjusted to 40:30:30 %, which is equivalent to 57:25:18 in vol.%. Lower resin contents are not appropriate, since the viscosity and flow of the compound would be too low and not all the particles can be wetted by the resin. The compound weight of 180 g in total was chosen for each of the net shape plates and 148 g for the non-structured top plate.

In the first step only the cavities of the steel mold (figure 3a) were filled and pre-compressed with a low pressure of about 1 MPa to avoid density gradients during the uni-axial pressing process. Then the remained powder was filled into the steel mold to obtain the base plate. The second densification process was initiated by choosing a maximum pressure of about 4 MPa. The densification of the compound was started at 80 ℃ where the resin powder lowered its viscosity and impregnated the particles. The curing of the plates inside the mold was performed at a maximum temperature of about 180 ℃ for 20 min. Due to the little shrinkage of the material within the cavities during the hardening of the resin the plate could be ejected very easily. The resin content of 30 % provided a good molding for all chosen fin geometries as can be observed in figure 3b. All the WBC plates showed a high stability and stiffness with the density of about 0.9 g/cm³ .

(a)   (b)

Figure 3. (a) Steel mold (216x105x4mm³) used for the net shape pressing and (b) net shape plate in the Wood Based Composite stage with fins in the mm-scale.

Plate to plate joining

The second process step is the plate to plate joining, which takes place directly after the molding process. It is advantageous to join the WBC instead of joining the C/C after the pyrolysis because the fins are sensitive to small cracks and might be damaged during the further processing. The prototypical 3D HX stack was produced by joining four similar WBC plates with fins to a flat plate on the top side as it is shown in figure 4a. In total 20 g of a commercial available wood glue (Ponal®) was used and distributed on the smooth surfaces homogeneously. The pure glue without any additives was used because the joint area should be free of any by-products after the high temperature treatment. The plates were laminated one after another and the maximum pressure of 4 MPa was applied after each bonding step. Finally, the surface of the stack was machined and the overlapping edges were cut away. The commercial wood glue provides for excellent joining due to the mechanical adhesion and the chemical bonding of the glue molecules with the particles of the WBC plate. In particular the chemical elements of the wood molecules like lignin and cellulose provide hydroxyl and polar groups, which can chemically bond. All these mechanisms result in sufficient adhesion of the boundary layers as can be seen in the joining model shown in figure 4b. The joining area was investigated in the C/C and C/SiSiC stage. It can be derived from figure 5 that the microstructure in the joint area is homogeneous and free of any defects after the pyrolysis as well as after the silicon melt infiltration.

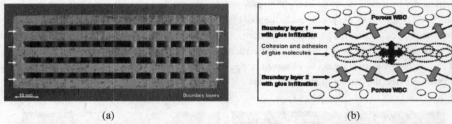

(a)   (b)

Figure 4. (a) HX stack after the joining process. (b) Joining model by using Wood Based Composites (WBC) as preforms and by using a wood glue as the binding agent.

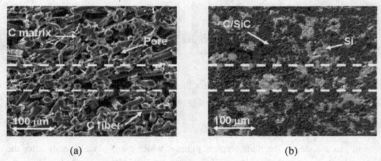

(a)   (b)

Figure 5. (a) Microstructure of the joint marked with dotted lines after the pyrolysis (C/C) and (b) after the silicon infiltration (C/SiSiC).

High Temperature Treatment (HTT) of the HX stack

The joined WBC derived HX stack was heated up to 1650 ℃ with a heating rate of about 2 K/min under nitrogen atmosphere to convert the organic based material into carbon/carbon (C/C). The shrinkage of the whole stack as well as of the different fin sizes must be low to avoid crack formation or de-bonding of the joints within the structure. The shrinkage of the fins with different length/width (l/w) ratios is shown in figure 6. It can be concluded from the graph that the length to width ratio of fins with different sizes remains nearly constant during the HTT. However, the higher the l/w ratio the higher is the deviation. The absolute shrinkage in the width is 4.7-11.7 % and slightly lower in the length which is about 4.2-7.8 %. It can be assumed that most of the carbon fibres are aligned parallel to the length of the fins and thus lead to a reduced shrinkage. During the HTT the mass of the HX stack 570 g (WBC) was reduced to 286 g (C/C), representing a mass loss of about 49.8 %. The dimensions were reduced from 161.5 x 116.7 x 43.6 mm³ to 155.1 x 112.0 x 36.0 mm³ so that the shrinkage can be calculated to about 17.4 % in the length and about 4 % in the height.

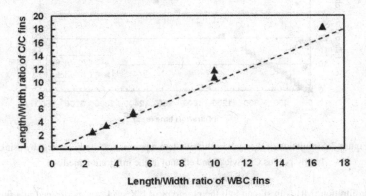

Figure 6. Shrinkage behavior of fins with various length-to-width ratios during pyrolysis at 1650 ℃.

Melt Infiltration (MI) of silicon into C/C by the Wick Method

Silicon melt is infiltrated into the porous C/C (porosity 68.8 vol.%) to convert the matrix carbon into SiC and to fill the residual porosity. However, the carbon fibres should be retained to obtain a carbon fibre reinforced C/SiSiC ceramic. The experiments must be conducted very carefully. On the one hand the blockage of the flow channels must be absolutely avoided. On the other hand the ceramic plates must be absolutely gastight (porosity ~0) to avoid leakage of the heat transfer fluid in the different layers under the required pressure differences.

In order to determine the best silicon infiltration parameters, the infiltration into C/C was simulated with ethanol ($\rho = 0.79$ g/cm³) as the reference medium at room temperature. The method and formulas used to calculate the filled pore volume from the up-take of ethanol (m = 16.32 g, V = 38.8 cm³, ρ = 0.57 g/cm³, P = 68.8 vol.-%) are described in detail by Hofenauer[8]. Porous wicks made of C/C (∅ = 10 mm, l = 10 mm, P~ 40 vol.%) were brought in direct contact with the C/C plate, which was adjusted in the horizontal position. The medium first soaks into the wicks and then into the plate. It can be seen in figure 7 that 6 wicks give better results compared with 3 wicks in regard to the total ethanol uptake. The infiltration with 6 wicks follows a linear trend and is actually not complete even after one hour. The low infiltration speed is favoured in order to avoid an overflow of silicon into the channels, which must remain open.

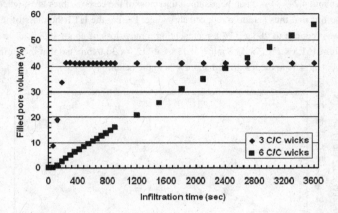

Figure 7. Simulation of the melt infiltration into a porous C/C body process by using highly porous C/C wicks and ethanol as the infiltration medium.

The melt infiltration (MI) with silicon into the prototypical HX stack was performed in a furnace with resistance heating at the maximum temperature of about 1650 ℃ under a pressure of about 1 mbar. The holding time at the maximum temperature was 45 min. Prior the MI some samples were cut from the C/C stack so that the mass was reduced to 257.45 g. The amount of the used silicon (grain ∅ 1-10 mm) was about ~978 g, which is about 3.8 times of the weight of the C/C. The mass gain of the HX stack was 346 % resulting in a total mass of about 1147.1 g. No measurable dimensional change could be detected after the MI process. The composition of the C/SiSiC was investigated by oxidation (800 ℃, 10 hours) followed by wet chemical analysis with HF/HNO₃ (1:1), and it was found out that the elemental composition is: SiC = 69 m.% (59 vol.%), Si = 19 m.% (22 vol.%), and C = 12 m.% (19 vol.%). The carbon fraction remained mostly in form of the carbon fibres. The outer surface of the ceramic HX was completely free of silicon drops as it is shown in figure 8. It can be clearly seen in micrographs in figure 9 that most of the carbon fibres were retained in the microstructure, and that the flow channels could be maintained open. However, there is evidence of some silicon droplets inside the volume, especially close to the flow channel. This should be attributed to the higher porosity in this region, leading to the higher silicon content in the C/SiSiC composite. It can be already detected in the WBC plates that the densification during the axial pressing in the fin volume is supposed to be lower. Due to the higher local porosity, the capillarity forces are lower. It can be assumed that the silicon melt cannot be retained in the microstructure and sweats out by forming some silicon drops on the surface. Also, in the corners where the original plates were bonded together, capillary forces create a small region of silicon with the shape of a radial fillet, as seen in figure 9b. This fillet may be helpful in reducing localized stress concentration in these joints.

Figure 8. Ceramic heat exchanger stack (155.1 x 112 x 36 mm³) made of C/SiSiC performed by plate to plate joining in the WBC stage, pyrolysis and final infiltration with silicon at 1650 ℃.

(a) (b)

Figure 9. (a) Microstructure of the carbon fibre reinforced C/SiSiC composite. Composite phases: SiC 59 vol.% (dark grey), silicon 22 vol.% (bright), carbon 19 vol.% (black) and (b) integrated flow channel within the HX stack.

CHARACTERISATION AND TESTING OF THE C/SiSiC CERAMIC COMPOSITES

Mechanical tests

Mechanical tests were performed with a Zwick 1475 testing apparatus (type testXpert) at room temperature by the 3 point bending strength test. The cross head speed was 0.5 mm/min, the preload was set to 10 N while the width of the stiff supports was chosen to 26.9 mm. The test geometry of the samples was rectangular and about 30.1 x 9.9 x 3.3 mm³. In total, a population of 24 samples were evaluated, which is a representative number to perform the statistical process. The samples were cut out from a single net shape plate with integrated flow channels in the ceramic stage. The fins were ground away and the surface of this base plate was polished. Half of the samples were cut from the ceramic panel in the direction parallel and half of the samples perpendicular to the fins in order to evaluate the materials' anisotropy. The scatter in data of strength values was evaluated with all samples by using the Weibull statistical process. This process takes non-homogeneities like pores, cracks or elemental agglomerations in brittle materials into account. The regression analysis fits the best straight line, which is fitted to a set of points. The regression was calculated by using the least square method (LSM). The flexural strength values were correlated with the microstructure as well with the composition of the C/SiSiC composite.

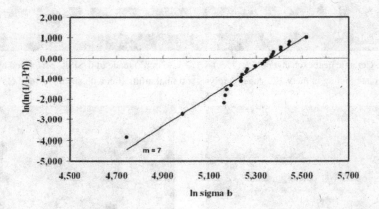

Figure 10. Weibull distribution of the strength values.
The linear regression line was calculated using the least-square method (LSM).

All samples experienced brittle fracture and showed a linear force-deflection response to the point of failure. Damage tolerance due to fibre reinforcement could not be expected due to the short length of the single fibres ($l \leq 400$ µm) and their low volume content (≤ 19 vol.%). The high brittleness can be confirmed by the low fracture toughness of 3.7 MPa√m, which was measured by the SEVNB method. The standard deviation of all flexural strength values was calculated to 29. The minimum strength was 115 MPa and the maximum strength was 252 MPa. The mean value was calculated to be 198 MPa. In figure 10 the failure probability Pf is plotted against the flexural strength σ_b values. The regression line has a linear form. The Weibull modulus 'm' was calculated to be 7. The reason for the low strength values can be probably due to inclusions like agglomerations of carbon fibres or large silicon grains within the microstructure.

Gas leakage testing

Helium permeation tests were performed on uncoated and CVD carbon coated C/SiSiC coupons (Ø = 50 mm, t = 3 mm) using a bubbles observation method [4]. The schematic of test design is shown in figure 11. A test coupon is clamped between two O-rings. High pressure helium is connected at the bottom side. The top side is covered with water. If there is any helium leakage through the test coupon, gas bubble will be observed. Before formal test, a copper disc was used to test hermeticity of the fixture. Under 0.5 MPa pressure difference, no visible bubble was detected after 10 minutes. This test verified that there was no baseline leak through O-rings. Different types of coupons were tested for hermeticity. Initially low pressure was applied. If no bubble observed, the helium pressure will be increased until that bubbles are observed.

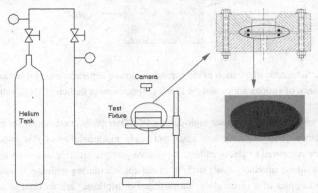

Figure 11. Helium permeation test and C/SiSiC test coupon (Ø = 50 mm, t = 3 mm) using a water bubble observing method.

The experiments show that uncoated C/SiSiC can have considerable permeability. The threshold pressure difference for detection of leaks is as low as 3.4 kPa. No visible bubbles appear for coated C/SiSiC coupons. The permeation test on the CVD SiC and pyrolytic carbon (PyC) coated samples show that the coating layer kept good hermeticity under high pressure up to 5.5 MPa. No helium bubbles were observed on the sample surface, even though some baseline leaks started to be detected from the fittings of the test fixture. The sample, however, failed about five minutes later. Subsequent calculation showed the sample was experiencing a tensile stress of 276 MPa before breaking, which is much higher than 40 MPa maximum tensile stress (under 9 MPa He) typically found from the stress analysis.

Experimental design for corrosion testing of uncoated and coated C/SiSiC composites in high temperature molten salt

Corrosion tests were performed in high purity, molten FLiNaK salt at 850 °C for 500 hours, in a graphite crucible. The alloy corrosion test results have been reported previously elsewhere [9]. Figure 12 shows a schematic illustration of the corrosion test capsule.

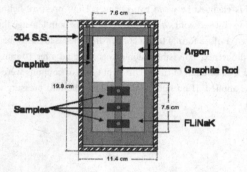

Figure 12. Schematic illustration of the corrosion capsule apparatus used for testing corrosion performance of various alloys and the C/SiSiC specimens in molten FLiNaK salt at 850°C.

The corrosion test capsule design uses high-density POCO graphite containment for molten salt rather than glassy carbon. The relatively low costs and ready machinability of this material made it an attractive choice compared to glassy carbon. A separate graphite crucible was used for each corrosion test making the minor amounts of salt sticking to the crucible during draining, a non-issue. The tests samples (three samples of a given alloy, alloys tested in triplicate) are fixtured to an axially placed central graphite rod using graphite screws. The dimensions of the C/SiSiC composite corrosion test samples (measured prior to corrosion testing) were as follows: the uncoated C/SiSiC composite was 12.95 x 31.83 x 1.41 mm³, the two coated samples were 13.20 x 32.24 x 1.78 mm³ and 13.15 x 32.18 x 1.79 mm³, respectively. The graphite crucible with the samples is placed in a stainless steel capsule for safety purposes and the top of the crucible is covered with graphite and stainless steel lids with a 1/4" opening for introducing the salt. The top stainless steel lid is then welded shut. All operations are carried out inside an argon atmosphere controlled glove box. After introducing the salt to the desired level, the opening is welded shut, this operation also being performed inside the glove box. The stainless steel capsule is then placed in a high temperature furnace for corrosion tests under argon atmosphere. After corrosion testing for 500 hours the furnace temperature is lowered to 500 ℃ and the test capsule is inverted in order for the salt to drain away from the samples. The stainless steel capsule is sectioned using a lathe and the samples are retrieved.

Pyrolytic graphite coated and uncoated C/SiSiC composites tested in FLiNaK

Pyrolytic carbon coated C/SiSiC composite is being considered for process heat exchanger material for intermediate salt loop. The C/SiSiC composites, 20 samples with $\rho = 2.51$-2.71 g/cm^3 and P = 0.1-3.8 vol. %), were fabricated and cut from flat plates by the German Aerospace Center, and the PyC coating was deposited by CVD process by Hyper-Therm High Temperature Composites, Inc., Huntington Beach, CA. The initial pre-corrosion characterization and post corrosion (in FLiNaK at 850 ℃ for 500 hours) of these samples was performed by SEM imaging of the cross-section of the samples and compositional line scan analysis using energy dispersive spectroscopy (EDS). Figure 13 shows a cross-sectional SEM image of the PyC-SiC coated C/SiSiC composite. The coating exhibits a dual structure (multilayer) with an outermost 2-4 μm pyrolytic carbon layer and the inner ~ 47 μm Si_xC_y layer. When combined with EDS line scan analysis (Figure 14) it is clear that the composite substrate itself consists of a Si_xC_y phase, as well as pure silicon and pure carbon phases representing either matrix or fibers. The ratio of C:Si in figure 14 is about 70:30, however owing to discrepancies in detection of lighter elements such as carbon by EDS it is speculated that the stoichiometry of this silicon carbide phase corresponds to SiC.

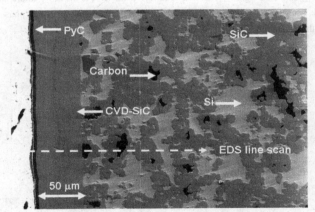

Figure 13. SEM cross sectional image of the C/SiSiC composite coated with outer pyrolytic carbon layer and inner SiC layer, both applied by the CVD process.

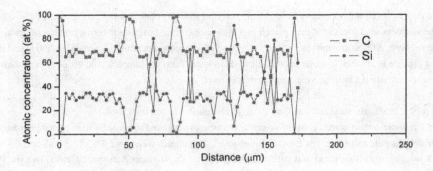

Figure 14. EDS line scan analysis across the C-SiC double layer coating and the C/SiSiC composite substrate, indicating phases like carbon, Si_aC_bSiC, and pure silicon.

Figure 15 shows the microstructure of the uncoated silicon-carbide composite after corrosion testing in FLiNaK at 850°C for 500 hours. The uncoated sample shows drastic attack of the microstructure along the grain and inter-phase boundaries. The corrosion attack appears to be particularly drastic for the Si phase, while the SiC and pure carbon phases are relatively less attacked by the molten salt. A qualitative understanding of the corrosion potential of the phases can be made by looking at the Gibb's free energy of fluoride formation per F_2 molecule (ΔG_{FF}) of the Si and SiC phases; the more negative the ΔG_{FF} for a particular phase, the more this phase can be expected to be attacked. At 850 °C, the ΔG_{FF} of SiF_4 is more negative than for SiF_2, it is therefore likely that SiF_4 is the more stable phase in FLiNaK salt. Furthermore, the ΔG_{FF} is more negative for pure Si than SiC, making the Si phase more prone to dissolution than the SiC phase.

Figure 15. Cross sectional SEM images of the uncoated C/SiSiC composite after exposure to molten FLiNaK 850 C/500 hours, showing severe corrosion attack along the grain and inter phase boundaries

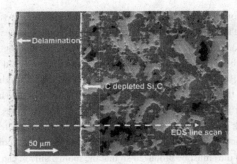

Figure 16. Cross sectional SEM images of the coated C/SiSiC composite after exposure to molten FLiNaK 850 ℃/500 hours with no remarkable attack.

Figure 16 shows the cross-sectional SEM image and figure 17 the EDS line scan of the coated C/SiSiC ceramic composite sample after exposure to molten FLiNaK at $850\,^{\circ}C$ for 500 hours. There is remarkably little attack of the coating, and consequently the base C/SiSiC composite after exposure to this aggressive environment. Based on figure 16, it appears that a carbon depleted Si_xC_y layer is formed as the interface between the CVD-SiC coating layer and the substrate, which may be responsible for the improved adhesion (white vertical line).

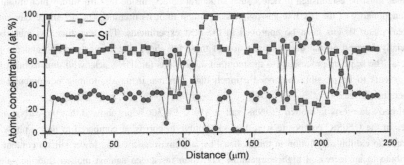

Figure 17. EDS line scan analysis across the coating and the coated C/SiSiC composite after corrosion test in molten FLiNaK at 850 ℃ for 500 hours. Distance on the horizontal axis corresponds to depth below surface.

CONCLUSIONS

Ceramic HX plates with integrated flow channels could be fabricated by the metal infiltration (MI) process with infiltrating molten silicon into highly porous C/C preforms. The molding and joining of first test plates derived from Wood Based Composites (WBC) seems to be a viable technology and has a very high potential for the up-scaling to large HX dimensions. Investigations with the HX test design provided good information, how the fins behaved, especially during the high temperature treatment. Even fins with low width of about 2 mm and length of about 10 mm could be fabricated with high reproducibility and are likely the favoured dimensions. Low length-to-width ratios of the fins < 10 should be selected, since they showed the lowest shrinkage rates <5 %. The shrinkage of the plate and the fins might be further reduced by increasing the amount of carbon fibres and reducing the wood powder content. The wick method is favourable for the net shape silicon infiltration into C/C preforms. However, very homogenous microstructures with a high carbon and SiC content as well as desired porosities (pore size < 20 µm) must be produced so that the molten silicon can be hold by the capillary forces within the bulk material during infiltration. The current silicon content of 22 vol.% and the grain size of up to $\varnothing = 50$ µm seem to be too high in regard to smooth surfaces without any silicon droplets. The scatter in data of the strength values and thus the Weibull modulus 'm' isn`t yet sufficient for a structural material and strongly depend on the uniformity of the C/SiSiC ceramic composite. The homogeneity must be increased by at least a factor of two. This can be probably reached by improving the net shape molding process. Microstructures with reduced density gradients are required, especially in the fin areas. Homogeneous WBC preforms can only be obtained by a highly accurate filling process and regular densification of the compound within the cavities. The pre-densification of the cavities or the bi-axial pressing process could be possible approaches to overcome this problem. Non-destructive methods must be established to detect such structural defects inside the structure, which might lower the functionality of the HX like gas-tightness or might limit its lifetime. The net shape fabrication of the real plate designs must be approved in the next experiments. The investigations must be then especially focused on the quality inspection of these net shape plates after pressing and the joining process. The scale-up processing to the required dimensions for IHX stacks with dimensions of about 1 m³ seems to be possible, but need a much high expense in molds, joining equipment and high temperature furnaces.

Initial corrosion tests in molten FLiNaK salt at 850 °C for 500 hours showed that the carbon and SiC phases in the C/SiSiC composite were less susceptible to corrosion compared to the Si phase which appeared to exhibit dissolution in the molten FLiNaK environment. The lower silicon content would contribute to an increased high temperature corrosion resistance against molten fluoride salts. The corrosion resistance of the composite was remarkably improved by the presence of a carbon and SiC bilayer coating, indicating the relative inertness of carbon and SiC phase in this environment. The bilayer coating did not exhibit any tendency to delaminate because of the formation of a Si-rich Si_xC_y layer at the coating-composite substrate interface. It appears that a coating would be necessary for use of this composite in molten FLiNaK salts at high temperature, however if the pure Si content of the composite is reduced a coating may not be required. Future tests are planned in molten chloride salts ($KCl-MgCl_2$ or $KCl-MgCl_2-LiCl$ salt) where it is speculated that the corrosion attack would be far less severe. Ultimately, the CVD coating process for the internal channels of compact ceramic heat exchangers must be confirmed, and corrosion testing performed in a heat transfer test loop. The work completed to date provides optimism that such compact ceramic heat exchangers could exhibit excellent performance.

REFERENCES

[1]Future Reactor Materials - A Revolutionary Reactor concept – ORNL Review Vol. 37, No.1, (2004).

[2]D. F. Williams, Assessment of candidate molten salt coolants for the NGNP/NHI heat-transfer loop, ORNTL-report/TM-2006/69, (2006).

[3]L.C. Olson, J. Ambrosek, K. Sridharan, M. Anderson, T. Allen, Y. Chen, M. Corradini, Evaluation of material corrosion in molten flouride salt, American Institute of Chemical Engineering Annual Meeting, San-Francisco, Nov. 12-17, (2006).

[4]P. F. Peterson, Haihua Zhao, Fenglei Niu, and Wensheng Huang, Jens Schmidt and Jan Schulte-Fischedick, Development of C-SiC ceramic compact plate heat exchangers for high temperature heat transfer applications, American Institute of Chemical Engineering Annual Meeting, San Francisco, Nov. 12-16, (2006).

[5]J. Schulte-Fischedick, V. Dreißigacker, R. Tamme, An innovative high temperature plate-fin heat exchanger for EFCC processes, Appl. Thermal Engineering 27, 1285-1294 (2007).

[5]V. Ponyavin, Y. Chen, M. B. Trabia, M. Wilson, A. E. Hechanova, Modelling and parametric study of a ceramic high temperature heat exchanger and chemical decomposer, Proc. of IMECE, Nov. 5-10, Chicago, (2006).

[6]J. Cutts, M. A. Wilson and V. Ponyavin, Dynamic flow of micro-channels in a ceramic heat exchanger, American Institute of Chemical Engineering Annual Meeting, Nov. 12-16, San Francisco, (2006).

[7]M. A. Wilson, C. Lewinsohn, J. Cutts and E. N. Wright, Optimizing of micro-channel features in a ceramic heat-exchanger for sulphuric acid decomposition, Proc. of AIChE conference, Nov. 12-17, San Francisco, (2006).

[8]A. F. Hofenauer, Entwicklung spezieller Holzwerkstoffe fü die Herstellung Silicium- infiltrierter Siliciumkarbid-Keramik; *Dissertation TU Müchen* , 315 p. (2004).

[9]L. C. Olson, J. W. Ambrosek, K. Sridharan, M. H. Anderson, T. R. Allen, Materials corrosion in molten LiF–NaF–KF salt, Journal of Fluorine Chemistry 130, 67–73, (2009).

MICROSTRUCTURAL STUDIES OF HOT PRESSED SILICON CARBIDE CERAMIC

Abhijit Ghosh, Abdul K. Gulnar, Ram K. Fotedar, Goutam K. Dey, Ashok K. Suri
Materials Group, Bhabha Atomic Research Centre
Mumbai, India

ABSTRACT

Silicon Carbide (SiC) ceramics belong to an important class of structural materials used for high temperature applications. The specific attributes that account for their utility as engineering ceramics are high value of thermal conductivity together with lower thermal expansion coefficient, thereby having higher thermal shock resistance. High mechanical strength also plays a vital role for considering SiC as structural components in various devices. However, it is difficult to densify SiC even at elevated temperature because of the covalent nature of its bond. In the present work, fine powder of β-SiC mixed with sintering additives (Al, B, C) was hot pressed at 32 MPa in graphite die at various temperatures in the range of 1700-1900°C under vacuum. Hot pressing at 1900°C yielded a fully dense material. Optical micrographs showed presence of elongated grain of α-SiC. Fracture surface of SiC observed under scanning electron microscope (SEM) revealed the occurrence of liquid phase sintering. The grain boundaries appeared to be free of secondary phase as revealed from the bright field conventional transmission electron microscopy (TEM) images. High-resolution electron microscopy analysis, however, provided evidence for distribution of amorphous phase at the grain triple junction. Hardness and toughness of the hot pressed SiC were found to be in the range of 20-22 $Hv_{0.1}$ and 2.5-3.0 $MPa.m^{1/2}$ respectively. Flexural strength of the highly dense samples was found to be in the range of 225-240MPa.

INTRODUCTION

In modern power installations, the chemical plant equipments and the metallurgical plant have joints and components which operate under conditions of aggressive media; high temperatures, cyclic thermal stresses and erosive wear conditions. Often, these components must also be stable in making rapid heat exchange possible. Hence, the materials having high strength, fracture toughness, creep resistance, thermal shock resistance and thermal conductivity at elevated temperature are the most preferred candidates to be used in such conditions. However, it is difficult to obtain all such properties in a single material. For example, silicon carbide is one of the candidate materials possessing all such properties except high fracture toughness[1]. Indeed SiC is one of the most important material in the family of non-oxide ceramics. It is also used as large scale structural components in very aggressive corrosive environments[2,3]. Due to its excellent mechanical properties at high temperature, this material finds application in rotors, vanes and blades in gas turbine, heat exchangers, ceramic fans, mechanical seals, pump shafts, bearing, sleeves, abrasive resistance liners etc[4-8]. Some of these applications have already been demonstrated while some others are in various stages of development.

However, optimum high temperature mechanical properties can only be realized in dense SiC ceramics with tailored microstructure. Typical fabrication processing employed for obtaining dense SiC include reaction bonding or self bonding process[9],chemical vapor infiltration, precursor impregnation and pyrolysis reactions. It has been demonstrated that the addition of secondary phases to SiC matrix can improve its density, mechanical properties, fracture toughness, flexural strength etc. Main additives used in the sintering of Si_3N_4 and SiC ceramics are mixtures of aluminum, boron-carbon, Al_2O_3-Y_2O_3, Al_2O_3-Gd_2O_3, Y_2O_3-AlN etc[10-15]. Incorporation of certain additives also helps in forming a favorable bimodal microstructure, which improves toughness of the material[16].

It has been demonstrated that densification of SiC can be carried out via liquid phase sintering mechanism at relatively low temperature (1850°C to 2000°C) by adding small quantities of sintering aids, viz., Al_2O_3[17] and Al_2O_3 + Y_2O_3 (or rare earth)[18-24]. The resulting materials have equiaxed fine-grain microstructures with second phases located at triple junctions of SiC grains[22,24,25]. Transmission electron microscopy (TEM) studies indicated absence of intergranular phases at the two grain junctions of SiC materials, thus making the material creep resistant[17]. Therefore, without sacrificing the high temperature properties, the sintering temperatures of SiC can be reduced by choosing liquid-phase sintering over solid state densification. However, the room temperature fracture toughness of the liquid-phase-sintered SiC materials remains low[21]. One of the strategies for toughening monolithic ceramics is to produce bimodal grain size distribution containing large, elongated grains in a matrix of smaller, equiaxed grains. Such a bimodal microstructure was developed by hot pressing SiC with Al, B, and C sintering additives (ABC-SiC) by Cao et al[26]. The Al, B, and C additives ensured the full densification of SiC,[26-29] while B and C also facilitated the low temperature β- α phase transformation during hot pressing[30-34]. Anisotropic growth rates of hexagonal α-SiC resulted in formation of platelike, elongated grains by consuming neighboring matrix grains[35].

In the present investigation, effort has been made to optimize the hot pressing temperature to obtain near-theoretically dense sample. Microstructural analysis of the fully dense sample was carried out and attempt has been made to correlate mechanical properties with the microstructural features.

EXPERIMENTAL

Fine powder of β-SiC (Electro Abrasives, USA) was mixed with the sintering additives, viz. Al (1.0%), B (0.5%) and C (3.0%). The carbon source in the present investigation was petroleum coke. The powders were mixed in a planetary mill for 1h in a polyamide lined pot. The wet mixing was carried out in alcohol using alumina balls as the mixing medium. The materials were dried at $80^{\circ}C$ in vacuum. Hot pressing was carried out at 32 MPa for 30 minutes in vacuum using graphite mould maintained at temperature in the range of 1700-1900°C.

Densities of the hot-pressed disks were measured by the Archimedes method. Test pieces of 3 mm × 4 mm × 36 mm were prepared from the disk by cutting and grinding using a diamond wheel and paste. These samples were used for measuring the flexural strength, Vickers hardness and fracture toughness. Flexural strength was measured by three-point-bending method over 30 mm span with a cross head speed of 0.1 mm/min. A Vickers diamond indenter with an indentation load of 0.98N was used for measuring hardness of the sample and a load of 9.8 N for generating diagonal crack for calculating indentation fracture toughness (K_{IC}). Data for hardness, flexural strength, and fracture toughness were collected on three to four specimens.

Microstructural development with temperature was examined by observing the polished and etched SiC sample under the optical microscope (magnification 500X). Murakami solution was used to etch the polished SiC sample. To understand the fracture mode in the fully dense sample, fracture surface of the sample was examined under the scanning electron microscope (SEM). To avoid the space charge effect during microscopic investigation, a thin layer of silver was coated on the surface of the sample.

Detailed microstructural and structural studies were conducted by transmission electron microscopic technique. For preparation of TEM sample, thin section of 300 μm was sliced from the pellets followed by ultrasonic drilling of 3 mm dia circular discs from it. Subsequently the disc was ground to ~ 100 μm thicknesses using a disc grinder and then thinned it down to ~35 μm in a dimple grinder. This sample was ion milled for electron transparency and finally coated with carbon to make surface conducting. A JEOL 200 FX microscope was used to image the general morphological features and a JEOL 3010 microscope with a point-to-point resolution of 0.21 nm was utilized to examine the intra and inter granular features (later one is known as high resolution transmission electron microscopy or HREM).

RESULTS AND DISCUSSION

Fine β SiC (0.5μ) powder in the presence of sintering additives i.e., Al, B and C were hot pressed within the temperature range of 1700- 1900°C in vacuum. A marked increase in density was observed as the processing temperature was increased from 1700 to 1900°C. Densification of the sample with the hot pressing temperature is shown in Fig.1.

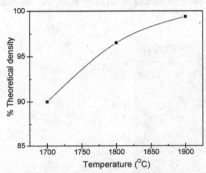

Fig.1. *Density of the hot pressed sample as a function of hot-pressing temperature.*

The addition of additives also helped in the conversion to high temperature α phase. Hot pressing involves simultaneous application of pressure and temperature resulting in an increase in particle mobility and contract stress within the ceramic powder thereby accelerating the densification kinetics. Inomata et al.,[36] reported formation of a liquid phase from Al_4C_3-B_4C-C system ~1800°C. In the present investigation, it is believed that formation of this liquid phase helped in reaching more than 96% TD (theoretical density) at and above 1800°C. Thus SiC having full density and fine grain sizes could be fabricated at lower temperatures and shorter cycle time than those required by conventional sintering.

Detailed microscopic studies have been conducted to determine the densification, grain growth and morphological behaviour. Typical optical micrographs are presented in Fig.2. It shows an increase in the grain size with temperature. It also indicates that the aspect ratio of the grain was increasing with increase in hot pressing temperature. The characteristic plate like morphology of α-SiC grains having

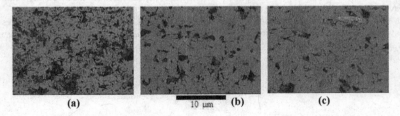

(a)   10 μm   (b)   (c)

Fig.2: *Optical micrographs of HP-SiC. The marker shows the general magnification. (a), (b) and (c) represent the sample hot pressed at 1700, 1800 and 1900 °C respectively.*

aspect ratio of about 3-5 was developed in the fully dense SiC sample i.e. sample hot pressed at 1900°C. The length and the width of the elongated grains were ranging from 5-15 μm and 1-4 μm respectively. Conversion to high temperature α-phase is generally very rapid at these temperatures in the presence of additives used[18]. In the present investigation, the grain size distribution is found to be bimodal in all the samples. The SEM fractograph of the 1900°C hot pressed sample reveals presence of intraganular fracture. The micrograph shows characteristics of a typical liquid phase sintered sample. The high density in the hot pressed sample is evident from the absence of pores present in the microstructure.

10 μm

Fig.3: SEM fractograph of 1900 °C hot pressed SiC.

The grains observed under TEM are shown in Fig.4. These are found to be exhibiting heavily faulted structure. In Fig.4 (a), the stacking faults extending up to grain boundary are observed (marked with A). The grain boundary (between B and D) seems to be quite clean. A triple junction between grain B, C and D is also observed. The inclusions are found to be present inside the grain and the position was close to the grain boundary. In presence of Al, B and C, SiC undergoes liquid phase sintering[26,30]. May be the unmixable liquid phase remains as inclusion within the grain. The size of inclusion is found to be within submicrometer range. The prominent features present in the microstructure are the polytypes of SiC, which are complex mixture of different variants. Presence of a heavily faulted structure is evident from the Fig. 4(a). In this figure, stacking fault is found to be running from the grain boundary to the interior of the grain. Termination of stacking fault within the grain may have created the dislocation network.

Multi grain structure in presence of inter-penetrating grains is shown in Fig.5. It indicates that grain A was sandwiched between grain C and D and was running under grain B. A comparison of grain A and B corroborates our earlier observation, i.e., presence of bimodal grain size in the near theoretically dense sample.

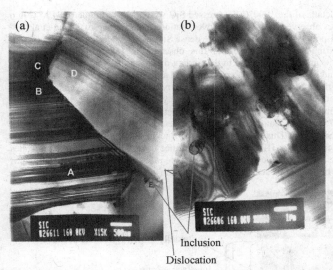

Inclusion

Dislocation

Fig.4: TEM micrographs of 1900°C hot pressed SiC. Features like inclusion, dislocation and stacking faults are shown here.

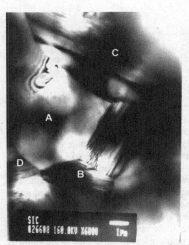

Fig.5. Multi grain structure observed under TEM.

Fine microstructural details of the 1900°C hot pressed SiC were observed under HREM (Fig.6 and 7). Comparing the features observed in the present investigation to the finding of Zhang et al.,[37,38] it has been confirmed that 4H-poly type was present in the dense SiC sample. Measured 'd' spacing of 0.25 nm in (0004) lattice -fringes also0 corroborates this observation. Interesting features were observed at the interface of SiC grains. As opposed to the finding of Zhang et al.,[37] the interface in the present investigation was found to be consisting of crystalline and amorphous phase. It is evident from Fig.6 that epitaxial crystallization at grain boundary (grain boundary adjacent to grain A) has lead to an epitaxial growth of a newly formed crystal (B) at the grain boundary. Most likely, this crystalline phase is α-Al$_2$O$_3$. Untransformed amorphous region can be observed between grain B and C. Thickness of this amorphous region is ~0.75nm. The feature like a zigzag structure present in the grain A and C is due to the polytopism in the SiC grain. This was formed due to stacking fault in the 3C-SiC structure. The triple grain junction observed under HREM revealed presence of amorphous phase. It indicates that discrete amorphous phase was mostly located at the grain triple junction. The increasing amount of zigzag structure was found to be present close to grain triple junction.

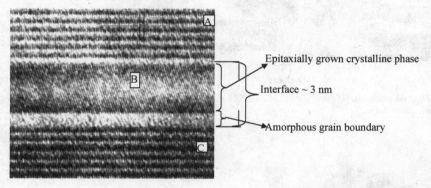

Fig.6: Grain boundary structure of SEM. The lattice parameter (marked with red color) was measured as 0.25 nm in (0004) lattice fringes.

Another interesting feature observed under the HREM was the presence of mirror like plane within the grain (Fig.8). Structure of this kind may be detrimental for mechanical properties, because the probability of formation of cleave like fracture always enhances in presence of mirror like plane (twin boundary). Infact, this may cause intragranular fracture in the sample. This kind of fracture reduces the strength and toughness of the material considerably. Heavily faulted structure is also found to be present within the grain, as evident from the Fig.8. From the microstructure, it looks like that; the dense SiC will not be having a very high toughness and flexural strength. The results of measured toughness and strength of the 1900°C hot pressed SiC are shown in Table-1. The value of toughness is found to be in the range of 2.5-3 MPa.m$^{1/2}$, despite the fact that there was formation of elongated inter-penetrating hexagonal grain in the matrix of SiC. This value is mostly reported in the literature.[21] However, for SiC sample sintered with Al, B and C as sintering aid and having microstructure similar to the present investigation, the reported toughness value were in the range of 6-9 MPa.m$^{1/2}$ [26,39,40]. In the present investigation, the flexural strength is also found to be quite low as compared to the value of ~600 MPa reported for the similar system[40]. These may be due to presence of twin related features along with high concentration of faulted structure within the grain. These may have caused the forma-

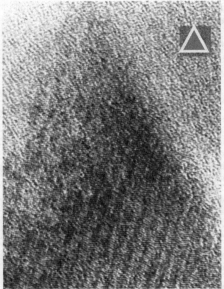

Fig.7. Crystalline grain with amorphous triple junction. Amorphous phase was marked with red arrow.

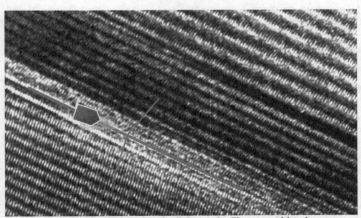

Fig.8. Presence of mirror plane in the SiC sample. The mirror like plane was marked with arrow head.

tion of intragranular fracture in the system, thus resulting in poor mechanical properties. Chen et al.,[40] observed intergranular fracture in their SiC sample causing high toughness and strength.

Table-1: Value of hardness, toughness and bend strength in hot pressed SiC.

Sample	Hardness ($Hv_{0.1}$)	Toughness ($MPa.m^{1/2}$)	Bend strength (MPa)
$1900^{O}C$ hot pressed	20-22	2.5-3.0	225-240

CONCLUSIONS

a. Hot pressing is a powerful technique to produce near theoretically dense SiC sample.
b. A combination of Al, B and C in small quantity acts as a very effective sintering aid in SiC.
c. The grains of SiC, after liquid phase sintering (hot pressed) shows faulty structure.
d. Epitaxial growth of crystalline phase in the grain boundary of SiC is evident; where as amorphous phase is found to be present at the triple grain junction.
e. Mechanical properties of the dense sample have not been improved because of the presence of intragranular fracture.

REFERENCE:

1. P.Chantikul, G.R.Anstis, B.R.Lawn and D.B.Marshall. "A critical evaluation of Indentation techniques for measuring fracture toughness:II strength method" J. Am. Ceram. Soc. 64 [9], 539-43(1981).

2. G.Magnani, G.L.Minoccari and L.Pilotti, "Flexural strength and toughness of liquid phase sintered silicon carbide" Ceram. Int. 26 [5], pp495-500(2000).

3. M.Steen and L.Ranzani, "Potential of SiC as a heat exchanger material in combined cycle plant" Ceram. Int. 26 [8],849-854(2000).

4. G.S.Upadhayay, "Sintered Metallic & ceramic Materials, John Wiley & Sons. pp.537-544(2000)

5. R.W.Cahn, P.Haasen & E.J.Kramer, Material Sciense and technology, Vol 17A,pp129-150(1996).

6. H.Hausner, Energy & Ceramics, Material Science Monographs, Ed. P. Vincenzini, Elsevier Scientific, Amsterdam, Vol.6, pp 582(1980)

7. R.Yuan, J.J.Kruzic, X.F.Zhang, L.C.De Jonghe, R.O.Ritchie, "Ambient to high temperature fracture toughness and cyclic fatigue behaviour in Al-containig silicon carbide ceramics" Acta Materialia, Vol 51[20]6477-6491,2003

8. X.Tong, T.Okano, T.Iseki, T.Yano, "Synthesis and high temperature mechanical properties of Ti3SiC/2SiC composites" Journal of Material Science, Vol 30[12],3087-3090,(1995).

9. C.W.Forrest, P.Kennedy, J.V.Shennan, Special ceramics, Ed. P. Popper, Vol.5.,pp99-123(1972)

10. S.Riberio, S.P.Taguchi,F.V.Molta, R.M.Baleslne, "The wettability of SiC ceramics by molten $E_2O_{3(ss)}$/AlN ($E_2O_{3(ss)}$=Solid Solution of Rare Earth Oxides)" Ceramics International, Vol. 33 (4),527-500(2007).

11. K.Biswas, G.Rixecker, L.Wiedmann, M.Schweizer, G.S.Upadhayay and F.Aldinger, "Liquid phase sintering and microstructural properties of SiC carbide ceramics with oxynitride additives" Mater Chem. Phys, 67[1-3] 180-191 (2001).

12. L. Wu, Y. Chen, Y. Jiang, and Z. Huang, "Liquid phase sintering of SiC with AlN-Re_2O_3 additives" Journal of the Chinese ceramic society,Vol36[5],593-596,(2008)

13. G.Arsalan and A. Kalemtas, "Processing of Silicon Carbide-boron carbide-Aluminium composites" ,Journal of the European ceramic society ,Vol29[3]473-480,2009

14. S.Prochazaka, Special Ceramics, Ed. P. Popper, Vol. 6,.171-84(1975)

15. M.Mader, F. Aldinger and M.J. Hoffman, "Influence of the α /β – SiC phase transformation on microstructural development and mechanical properties of liquid phase sintered Silicon Carbide", Journal of material Science, 34[6]1197-1204(1999)

16. U.K.patent 1,556,173.

17. K.Suzuki and M.Sasaki, "Pressureless Sintering of Silicon Carbide"; in Fundamental structural ceramics Edited by S.Somiya and R.C.Bradt, Terra Scientific Publishing Company, Tokyo, Japan, pp75-87, (1987).

18. M.Omori and H.Takei, "Pressurless sintering of SiC," J.Am.Ceram.Soc.,65[6]C-92(1982).

19. M.Omori and H.Takei, "Composite Silicon Carbide Sintered Shapes and its manufacturing," U.S.Pat.No.4502983, (1985).

20. M.Omori and H.Takei,"Method for preparing Sintered Shapes of Silicon Carbide,U.S.Pat.No.4564490,(1986).

21. R.A.Cutler and T.B.Jackson,"Liquid Phase Sintered Silicon Carbide"; in ceramic Materials and Components for Engines, Proceedings of the Third International Symposium. Edited by V.J.Tennery. The American Ceramic Society, Westerville, OH, pp.309-18 (1989).

22. L.Cordery, D.E.Niesz, and D.J.Shanefield,"Sintering of Silicon Carbide with Rare-Earth Oxide additions" in Sintering of advanced ceramics, Vol.7.Edited by C.A.Handwerker, J.E.Blendell, and W.A.Kaysser. The American Ceramic Society,Westerville, OH, ;pp.618-36 (1990).

23. M.A.Mulla and V.D.Krstic, "Low Temperature Pressureless sintering of β-Silicon Carbide with Aluminium Oxide and Yttrium Oxide Additions", Am.Ceram.Soc.Bull.,70[3]439-43(1991).

24. D.H.Kim, C.W.Jang, B.H.Park, and S.G.Baik, "Pressureless-Sintering of Silicon Carbide with Additions of Yttria and Alumina," J.Korean.Ceram.Soc.,26[2]228-34(1989).

25. L.S.Sigi and H.J.Kleebe,"Core/Rim Structure of Liquid-Phase-Sintered Silicon Carbide,"J.Am.Ceram.Soc.,76[3]773-76(1993).

26. J.J.Cao, W.J.MoberlyChan, L.C.De Jonghe, C.J.Gilbert, and R.O.Ritchie, "In Situ Toughened Silicon Carbide with Al-B-C Additions." J.Am.Ceram.Soc.,79[2]461-69 (1996).

27. R.A.Alliegro, L.B.Coffin, and J.R.Tinklepaugh, "Pressure-Sintered Silicon Carbide," J.Am.Ceram.Soc.,39[11] 386-89 (1956).

28. S.Prochazka and R.M.Scanlan, "Effect of Boron and Carbon on Sintering of SiC," J.Am.Ceram.Soc.,58[1-2]72 (1975).

29. H.Tanaka, Y.Inomata, K.Hara, and H.Hasegawa, "Nnormal Sintering og Al-Doped β-SiC." J.Mater.Sci.Lett.,4.315-17(1986).

30. B.W.Lin, M.Imai , T.Yano, and T.Iseki, "Hot-Pressing of β-SiCpowder with Al-B-C Additives,"J.Am.Ceram.Soc.,69[4]C-67-C-68(1986).

31. A.H.Heuer, G.A.Fryburg, L.U.Ogbuji, and T.E.Mitchell, "β-α transformation in Polycrystalline SiC: I,Microstructural Aspects," J.Am.Ceram.Soc.,61[9-10]406-12(1978).

32. T.E.Mitchell, L.U.Ogbuji,and A.H.Heuer, "β- α transformation in Polycrystalline SiC: II. Interfacial Energetics," J.Am.Ceram.Soc.61 [9-10]412-13(1978).

33. L.U.Ogbuji,T.E.Mitchell, and A.H.Heuer, "β-α transformation in Polycrystalline SiC: III. The thickening of α Plates," J.Am.Ceram.Soc., 64[2] 91-99 (1981).

34. L.U.Ogbuji,T.E.Mitchell, A.H.Heuer and S.Shinozaki, "The.β-α transformation in Polycrystalline SiC: IV., "A comparison of Conventially Sintered Hot-Pressed, Reaction Sintered, and Chemically Vapour Deposited Samples." J.Am.Ceram.Soc., 64[2] 100-105 (1981).

35. W.J.MoberlyChan, J.J.Cao, and L.C.De Jonghe, "The Role of the Amorphous grain Boundaries and the β-α transformation in Toughened Sic," Acta Mater.,46{5}1625-35 (1998).

36. Y. Inomata, H.Tanaka, Z.Inoue and H.Kawabata, "Phase relation in SiC-Al₄C₃-B₄C system at 1800°C" J.Ceram.Soc.Japan (Yogyo Kyokaishi) 88 [6] 353-55 (1980).

37. X.F.Zhang, M.E.Sixta and L.C.De Jonghe., " Grain boundary evolution in hot pressed ABC-SiC" J.Am.Ceram.Soc. 83[11] 2813-20 (2000).

38. X.F.Zhang, M.E.Sixta and L.C.De Jonghe., " Secondary phases in the hot pressed aluminium-boron-carbon-silicon carbide" J.Am.Ceram.Soc. 84[4] 813-20 (2001).

39. D.Chen, C.J.Gilbert, X.F.Zhang and R.O.Ritchie., "High temperature cyclic fatigue-crack growth behaviour in an insitu toughened silicon carbide" Acta Mater. 48[3] 659-74 (2000).

40. D.Chen, X.F.Zhang and R.O.Ritchie., "Effects of grain boundary structure on the strength, toughness, and cyclic-fatigue properties of a monolithic silicon carbide" J.Am.Ceram.Soc. 83[8] 2079-81 (2000).

DIFFUSION BONDING OF SILICON CARBIDE TO FERRITIC STEEL

Zhihong Zhong [*,1)], Tatsuya Hinoki[2)], Akira Kohyama[2)]

(1) Graduate School of Energy Science Kyoto University, Gokasho, Uji, Kyoto 611-0011, Japan

(2) Institute of Advanced Energy, Kyoto University, Gokasho, Uji, Kyoto 611-0011, Japan

ABSTRACT

For SiC to be used either as structural materials or as functional components, solid-state interfacial reactions of SiC with selected metals are frequently encountered. In this work, the solid-state reactions of SiC with ferritic steel containing primary alloying elements Fe and Cr were investigated and joining of SiC to steel by inserted interlayer was presented. Intensive interfacial reaction was generated in the reaction zone. The carbon precipitates were embedded in a matrix of metal silicide. Cr reacted with C to form chromium carbide. Cracks in SiC substrate were observed due to the large residual stress which caused by coefficient of thermal expansion (CTE) mismatch. A tungsten compound as interlayer between SiC and steel for residual stress relaxation was proposed. Two step diffusion bonding procedure, bonding of SiC to tungsten compound at 1350 ℃ and subsequent joining to steel at 900 ℃, was carried out. Microstructural evaluation of the interfaces revealed that the interface bonded well. The shear strength of the joints and hardness across the interface were also elevated. Failure during shear strength test was occurred along the interface between SiC and tungsten compound.

1. INTRODUCTION

Silicon carbide (SiC) is one of the most promising high temperature materials for structural and functional application in nuclear and aerospace field due to its attractive mechanical properties and resistance to high temperature corrosion. The low fracture toughness and high cost machining, however, retarding its massive practical application in engineering where requires large and complex shape ceramic components. Although the technology for developing the silicon carbide fiber reinforced SiC matrix composite (SiC$_f$/SiC composite) could alleviate this suffering to some extent, fabricating of the large sizes or complicated shapes is still difficult and expensive. Another promising alternative is to combine SiC and SiC$_f$/SiC composite with metallic component or structure, generally steel or high temperature strength resistance alloys. In spite of many techniques have been developed for joining of ceramics to metals, only little attempts for SiC/steel system has been made during the last several decades, to our knowledge, and most of them were focus on clarifying the interfacial characterization [1-4]. In this work, the solid-state reactions of SiC with ferritic steel containing primary alloying elements Fe and Cr were investigated and joining of SiC to steel by inserted interlayer was presented. In addition, as both of SiC fiber and SiC matrix could be simulated by SiC bulk material in the case of SiC$_f$/SiC composite, the study of joining SiC ceramic to steel is also contribute to the development of the joining technique of SiC$_f$/SiC composite to steel.

Joining of SiC with steel by conventional fusion welding is undesirable because SiC vaporizes rather than melts at elevated temperature [5]. Furthermore, residual stress may develop during cooling in the joints mainly due to the large coefficient of thermal expansion (CTE) and elastic modulus mismatch between SiC and steel. In order to reduce the residual stress in the SiC/steel joints, some researchers have successfully used different interlayers between SiC and steel to minimize or reduce the residual stress in the joints. Yano et al. [1] used Ti/Mo as insert material whereas Naka et al. [3] tried Nb as interlayer to reduce the residual stress in the joints. In this work, a tungsten compound used as interlayer between SiC and ferritic steel (FS) for relax the residual stress was developed.

2. EXPERIMENTAL
2.1. The fabrication and annealing treatment of SiC/FS diffusion couple

The commercially available Hexoloy SA α-SiC substrate, whose density is about 99.5 % of the theoretical value, were used. They were formed by sintering at temperature exceeding 2000 ℃. Prior to diffusion bonding, the SiC plate were ground and cut into small pieces with dimension of 22 × 22 ×2 mm. The bonding surfaces were polished up to 1   μm, followed by ultrasonic cleaning in acetone and finally dried in air. The ferritic steel (FS, containing 0.07 wt. % C) has an approximate chemical composition of Cr-17, Fe-balance (in wt. %), and small amount of Si, Mn and Ni. The metallic part was cut from bulk material into 2.0 mm thick with 22 ×22 mm in dimension and mechanically ground on emery paper up to 1500 grit.

The assembled SiC/FS diffusion couples were annealed in a vacuum hot-pressing furnace, under a pressure of 5~10 MPa at 1000 ℃ for in a range from 30 to 90 min. The temperature was raised to 550 ℃ at 5 ℃/min and then heated up to bonding temperature with a rate of 10 ℃/min; the cooling rate was 5 ℃/min to 400 ℃ and followed by the samples were allowed to cool in vacuum. The vacuum in the furnace was kept below 10^{-2} Pa during the annealing cycle.

2.2. Diffusion bonding of SiC to FS with W-Cr-Ni

To lower the joining temperature and to obtain a controllable CTE of tungsten compound, small mount of chromium (Cr) powder and nickel (Ni) powder used as sintering additive were added into the tungsten (W) powder, dues to the suitable eutectic point (1345 ℃) for Cr-44 at. % Ni alloy based on the binary Cr-Ni phase diagram, and Cr will not form intermetallic compound with W while Ni is a common binder for tungsten sintering. The tungsten compound (W-Cr-Ni) was composed of W powder (purity: > 99.9 %, average particle size: 0.6 μm), Cr powder (purity: 99.9 %, average particle size: 63 μm) and Ni powder (purity: 99.9 %, average particle size: 2 μm). W, Cr and Ni powders with the weigh ratio of 97.9:1.2:0.9 were mixed in agate for 30 min. After mixing, the mixture was weighed according to the W-Cr-Ni thickness requirement. Then the mixture was put on the polished SiC surface in a graphite mould. The bonding pressure, 20 MPa, was applied on the specimen. Bonding experiments were performed in hot-pressing furnace under the flowing argon at 1350 ℃ for 1 h.

Before joining to steel, the facing surface of hot-pressed SiC/W-Cr-Ni was ground and polished by the emery paper #1500. The steel was treated the same as described in section 2.1. The couples were diffusion bonded in vacuum, at 900 ℃ for 30 min, under 10 MPa.

2.3 Microstructure observation and mechanical strength test

Following the diffusion bonding, the cross-sections of the diffusion zones in SiC/FS and SiC/W-Cr-Ni/FS were prepared for metallographic examination by standard polishing techniques up to 1 μm. The microstructure was examined by field-emission scanning electron microscope (FE-SEM, JEOL6700) with energy dispersive X-ray spectrometry (EDS). The concentration profiles of Fe, Cr, Ni and W were measured along the W-Cr-Ni/FS by an electron probe microanalyses (EPMA). The hardness along the bonded interface was evaluated by Nano-Indenter (ENT-1100a). The shear strength of SiC/W-Cr-Ni/FS joints were determined by a specially testing apparatus (Fig. 1) using sub-size specimens (3 × 3× 5 mm) which cut from the samples. Specimens were inserted in the testing apparatus and the testing apparatus was placed on an Instron tensile machine. Tests were carried out at room temperature and at a crosshead speed of 0.5 mm/min. Five samples were tested for each holding time.

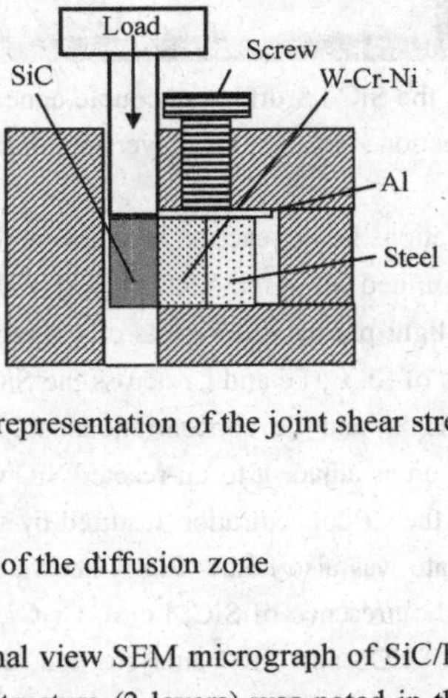

Fig.1. Schematic representation of the joint shear strength testing

3. RESULTS AND DISCUSSIONS

3.1. Microstructural characterization of the diffusion zone

3.1.1 Reactions between SiC and FS

Fig. 2 shows a cross-sectional view SEM micrograph of SiC/FS diffusion couple annealed at 1000 ℃ for 60 min. Layered microstructure (2 layers) was noted in the reaction zone. Immediately adjacent to the un-reacted SiC component is a multi-phase layer (layer 1), is followed by a light gray-colored layer (layer 2), and then the un-reacted steel substrate. It was found that the thickness of the reaction layers increased with extending the annealing time. The microstructure and phase composition of the reaction product, however, were independent on the annealing time. Many precipitates, randomly embedded in layer 1, were also noted and which were identified by EDS microanalysis to be 93.42C-0.45Si-0.41Cr-5.72Fe (in at. %). Hence, it can be inferred that the possible main phase of the precipitates is carbon (C) and was confirmed by after mentioned XRD analysis. Moreover, it is interesting to note that the sizes and volume fraction in certain area of C precipitates increased as their location further away from the SiC reaction front. In contrast, no precipitates were detected in layer 2 which appears as a clean layer. Furthermore, clustered voids were noted between layer 2 and FS. Cracks were found in SiC near to the layer 1.

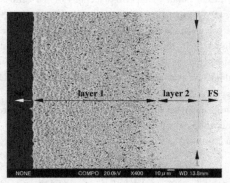

Fig. 2. Cross-sectional image of the SiC/FS diffusion couple annealed at 1000 ℃ for 60 min, showing the formation of two-layered reaction zone. The two layers are indicated as layer 1 and layer 2.

The elemental concentrations in the reaction layers were determined by EDS and the phases of the reaction products were determined using the XRD analysis. The phase composition analyses were only performed on the gray or light phases, the results of C precipitates were not given here. Fig. 3 shows the concentration profiles of Si, C, Fe and Cr across the SiC/FS reaction zone corresponding to the tested positions. In layer 1, Si and Fe concentrations are almost of constant (with Fe more pronounced than Si), except the areas adjacent to un-reacted SiC where having high Cr concentration. It can be noticed, however, that the Cr concentration featured by somewhat fluctuating in layer 2. The penetration of Si into FS substrate was also noted. Fig. 4 shows the XRD pattern of the bulk SiC/FS reaction zone, which indicates the presence of SiC, Fe_3Si, Cr_7C_3, C and a solid solution of Si in FS. Taking into account the results of EDS and XRD analyses that, the matrix phase in layer 1 should be the Fe_3Si which dissolved small amount of Cr and C. It has been reported [6] that, the Fe_3Si phase has a solubility of about 17 at. % Cr, which is much higher than that of the present detected value. The C precipitates in layer 1, which were the SiC decomposition product at elevated temperature during the SiC/FS reaction, were also detected by XRD. The chemical inertness between C and various metal silicides has been widely found [4, 7-9] during the interfacial reaction between non-strong carbide formation metals (alloys) and SiC. The reason why the dissociated C atoms formed precipitates is mainly attributed to the minimization of the total interfacial energy between the C atoms and metal silicides, from the point of view requirement of equilibrium system.

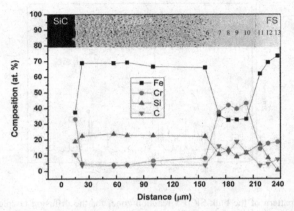

Fig. 3. Phase composition (at. %) and the corresponding analyzed points in the reaction layers formed at 1000 ℃ for 60 min.

Only Fe_3Si phase formation but no carbides were found in layer 1, suggesting that Fe has a stronger reaction affinity for Si than for C, which in turn gives rise to the reaction between Fe and Si and generated a Fe_3Si silicide matrix, at the temperature of interest. Thermodynamic data [7] show that the formation enthalpies of Fe_3Si and Fe_3C are negative and positive, respectively. Therefore, the formation of Fe_3Si, not Fe_3C, was thermodynamically favored. Although there exist several other Fe-silicides and Cr-silicides in Fe–Si and Cr-Si binary systems [6], thermodynamic considerations [7] show that Fe_3Si is much more stable than the other possible metal silicides. Thus it is reasonable and not surprising to find Fe_3Si phase in the reaction zone.

On the other hand, it is clearly seen that, from layer 1 to layer 2, the Fe concentration decreases from 69 to 33 at. %, while the Cr concentration increases from 6 to 40 at. %. Correspondingly, the Si and C concentration decreased and increased, respectively. Considering the results obtained from EDS and XRD analyses, the layer 2 is believed to mainly consist of Fe_3Si and Cr_7C_3. $Cr_{23}C_6$ may be present according to the EDS analysis, but its content might be too low to be detected. Surprising high Cr concentration was found in layer 2 compared to other region even in the FS, is probably due to the following reasons: (1) a slower diffusion (or rejection) rate of Cr as compared to Fe diffusion rate; (2) a supply of an extra amount of Cr atoms from the reaction, which segregated and formed Cr-rich layer and (3) prevailing selective reaction of Cr with C, rather than with Si, and formed Cr carbides in layer 2. A similar phenomenon has been reported [10], in which a metal-carbide layer containing Cr, Fe, Si and C was observed in the SiC/Fe-Cr alloy solid state reaction.

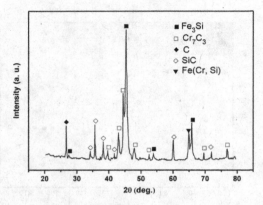

Fig. 4. The XRD pattern of the bulk SiC/FS reaction zone, for the diffusion couple annealed for 60 min.

It is well known that SiC is non-stable when it contacts with most of metals at elevated temperature. During the annealing process, Fe and Cr atoms diffuse from steel toward SiC and reacted with it; at the same time, the decomposed Si and C atoms diffuse to FS, and formed the reaction product. Fe reacted with SiC to form Fe_3Si and C following the equation:

$$SiC + 3\,Fe \rightarrow Fe_3Si + C \tag{1}$$

Simultaneously, Cr reacted with C to form Cr_7C_3, which is expressed by

$$7\,Cr + 3\,C \rightarrow Cr_7C_3 \tag{2}$$

However, the amount of decomposed C was much larger than that consumed by equation (2). This was the reason also why the C precipitates remained in the reaction zone.

In order to decompose SiC, a driving force is required to overcome the activation barrier of decomposition of SiC, which is thermally stable at the temperature investigated [11]. In addition, the C precipitates in the Fe_3Si silicide matrix also requires the formation of both interfacial and strain energies. Considering the overall reaction between SiC and steel, $SiC + FS \rightarrow [Fe_3Si + C] + Cr_7C_3 + \alpha\text{-}Fe(Cr, Si)$, thermodynamically, the driving force is believed to originate primarily from the highly negative total Gibbs free energy (ΔG) of the metal silicides and carbides formation. However, the ΔG

values are difficult to estimate due to the complicated chemical composition of the metal silicides and carbides.

3.1.2 Characterization of SiC/W-Cr-Ni/FS bonding zone

The cross-sectional view of SiC/W-Cr-Ni/FS joint was shown in Fig. 5, where the reaction layers formation were not observed, neither in the SiC/W-Cr-Ni nor W-Cr-Ni/FS interface. Some micro-sized pores were noted in tungsten compound where near to the SiC/W-Cr-Ni interface as well as in W-Cr-Ni interlayer. It is worth to mention that Ni-Cr (the gray phases) in tungsten compound tends to agglomerate in the SiC/W-Cr-Ni interface (Fig. 6). Furthermore, the Cr in the W-Cr-Ni interlayer was also found to be not finely dispersed, and near around the high Cr concentration region, it seems that the pore formation was readily, as indicated in Fig. 5(b). This is may be due to the large particle size of Cr (63 μm) and the un-well dispersed powder mixture. On the other hand, the amount of Cr and Ni additive in W was too small to produce enough Ni-Cr eutectic phases to fill the pores. The element concentration distributions of W, Ni Cr and Fe across the W-Cr-Ni/FS interface (along the red line) show the sharp in nature, which demonstrates no reaction phase formation but well bonding.

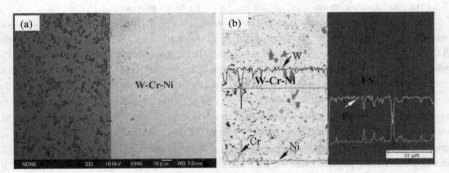

Fig. 5. The cross-sectional view of SiC/W-Cr-Ni/FS joint, (a) SiC/W-Cr-Ni interface and (b) W-Cr-Ni/FS interface.

Fig. 6. Enlarged image in Fig. 4(a), showing the Ni-Cr tends to segregate in the SiC/W-Cr-Ni interface and micro-pores.

3.2. Mechanical properties evaluation of SiC/W-Cr-Ni/FS joints

3.2.1 Hardness distribution along the SiC/FS and SiC/W-Cr-Ni/FS interfaces

Fig. 7 shows the hardness measurement across the SiC/FS and SiC/W-Cr-Ni/FS interface, respectively. The hardness for SiC is high and drops drastically as it crosses the interface. For the SiC/FS diffusion couple, the matrix of reaction layer 1, Fe_3Si, whose hardness value is about 5.3 GPa, which is similar to the reported value [7]. Passing through the layer 1, the hardness increases again to about 11 GPa in layer 2. The hardness increases gradually from layer 1 to layer 2, with increasing the content of Cr_7C_3. Due to the diffusion of silicon into the steel, the hardness near to the layer 2 can be seen to be clearly higher than that of the other region in steel. Similarly, the hardness in Fe(Cr, Si) diffusion layer decreases with decreasing the Si content. The hardness in steel from Fe(Cr, Si) about 25 μm is higher than that of farther regions is noted. It is probably due to the formation of σ phases [12] during the annealing process. However, the hardness of the farther region, about 4.5 GPa, is equal to that of the annealed this steel and indicates that no phase transformation has taken place during the annealing process. For the SiC/W-Cr-Ni/FS joints, the hardness of W-Cr-Ni is about 7 GPa, while the hardness value of steel is comparable to that in SiC/FS joints. Since only limited interdiffusion and not intensive interfacial reaction occurred, the hardness in W-Cr-Ni near to the SiC and the hardness near to the FS were not changed greatly.

3.2.2 Shear strength

The shear strength test was carried out on the SiC/W-Cr-Ni and SiC/W-Cr-Ni/FS joints. Typical plots of shear stress versus crosshead displacement for the two joints are displayed in Fig. 8. For the SiC/W-Cr-Ni joints, the average shear strength is 43.7 MPa. The failure occurred along the

SiC/W-Cr-Ni interface. Some small SiC flakes were attached on the fractured surfaces. For the SiC/W-Cr-Ni/FS joints, the average shear strength is 21.7 MPa. The failure was also happened along the SiC/W-Cr-Ni interface. Although the residual stress caused by CTE mismatch in SiC/W-Cr-Ni/FS joints has not been calculated by modeling or experimentally measured by X-ray diffraction method, it was believed to responsible for the strength reduction. Since the SiC/FS joints can not be machined for shear strength test (Crack was found in SiC after joining), it is obvious that the residual stress in SiC/W-Cr-Ni/FS joints was relaxed partly by the W-Cr-Ni interlayer. The shear strength of the SiC/Ti/Mo/steel joints brazed by Ag-Cu-Ti alloy [1], was about 30 MPa. Naka et al. [3] reported that SiC couldn't be joined to stainless steel when the niobium interlayer thickness was smaller than 1 mm. The shear strength increased to about 50 MPa when the joints fabricated by 4 mm thick niobium interlayer. In this work, the SiC/W-Cr-Ni/FS joints with appropriate strength were produced by using a simple and flexible powder metallurgy method. It is expected that such joints have a potential to be used at high temperature.

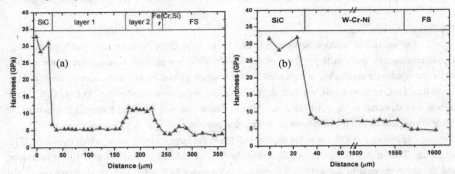

Fig. 7. Hardness profiles across the (a) SiC/FS interface and (b) SiC/W-Cr-Ni/FS interface, for the joints bonded at 1000 ℃ for 30 min, and   1300 ℃/ 60 min subsequent 900 ℃/ 30 min, respectively.

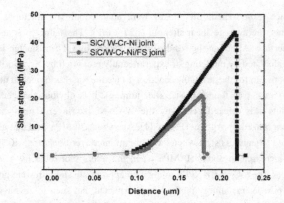

Fig. 8. Shear strength versus crosshead displacement for the SiC/W-Cr-Ni and SiC/W-Cr-Ni/FS joints.

4. CONCLUSION

The interfacial reaction between SiC and FS at 1000 ℃ for different time has been studied. The microstructure and reaction phases of the SiC/FS diffusion couples were similar to each other, while the reaction zone thickness depended on the annealing time. Fe has a stronger reaction affinity for Si than for C to form Fe_3Si, whereas the Cr prevailing selective reacted with C to form Cr_7C_3. Free carbon was detected in the reaction zone. Si also penetrated into FS. The hardness test across the reaction zone confirmed the formation of the reaction product.

The cracks in SiC/FS couples were noted due to the large residual stress which caused by CTE mismatch. The SiC/FS joints were successfully obtained by using a 1.5 mm thick W-Cr-Ni interlayer, and the shear strength of the joint was measured to be about 21.7 MPa. Neither intensive interfacial reaction nor strong interdiffusions were occurred in the SiC/W-Cr-Ni/FS joints.

ACKNOWLEDGEMENT

The authors are grateful to the finance support from National Institute for Fusion Science (NIFS), Japan.

FOOTNOTES
*Corresponding Author: TEL:+81-774-38-3465; FAX:+81-774-38-3467. E-mail Address:zh_zhong@iae.kyoto-u.ac.jp

REFERENCES
[1]T.Yano, N.Takada, and T.Iseki, Joining of pressureless-sintered SiC to stainless steel with Ag-Cu alloy

and insert metals, *The Ceram. Japan,* **95-3**, 357-382 (1987). In Japanese.

[2]T.C. Chou and A. Joshi, Selectivity of Silicon Carbide/Stainless Steel Solid-state Reactions and Discontinuous Decomposition of Silicon Carbide, *J. Am. Ceram. Soc.,* **74-6**, 1364-1372 (1991).

[3]M. Naka and T. Saito, Niobium interlayer for joining SiC to stainless steel, *J. Mater. Sci. Lett.,* **10**, 339-340 (1991).

[4]R. C. J. Schiepers, J. A. van Beek, F. J. J. van Loo and G. de With, The Interaction between SiC and Ni, Fe, (Fe,Ni) and steel: Morphology and Kinetics, *J. Euro. Ceram. Soc.,* **11**, 211-218 (1993).

[5]T.J.Moore, Feasibility Study of the Welding of SiC: *J. Am. Ceram. Soc.,* **68-6**, c151-c153 (1985).

[6]M. Lindholm, A thermodynamic description of he Fe-Cr-Si system with emphasis on the equilibria of the sigma (σ) phase, *J. Phase Equilibria,* **18**, 432-440 (1997).

[7]W.M. Tang, Z.X. Zheng, H.F. Ding and Z.H. Jin, A study of the solid state reaction between silicon carbide and iron, *Mater. Chem. Phys.,* **74**, 258–264 (2002).

[8]T.C. Chou, A. Joshi and J. Wadsworth, Solid state reactions of SiC with Co, Ni and Pt, *J. Mater. Res.* **4**, 796–809 (1991).

[9]K. Bhanumurthy and R. Schmid-Fetzer, Interface reactions between silicon carbide and metals (Ni, Cr, Pd, Zr), *Composites: Part A,* **32**, 569-574 (2001).

[10]M. B. Ricoult, Solid state reactions between silicon carbide and (Fe, Ni, Cr) alloys: reaction paths, kinetics and morphology, *Acta metal. mater.,* **40**, Suppl., s95-s103 (1992).

[11]T. C. Chou and T. G. Nieh, Solid state reactions between Ni_3Al and SiC, *J. Mater. Res.,* **5**, 1985-1994 (1990).

[12]M. Ghosh, Samar Das, P.S. Banarjee and S. Chatterjee, Variation in the reaction zone and its effects on the strength of diffusion bonded titanium–stainless steel couple, *Mater. Sci. Eng. A,* **390**, 217–226 (2005).

Ceramics for Fuel Coating

FRACTURE PROPERTIES OF SiC LAYER IN TRISO-COATED FUEL PARTICLES

Thak Sang Byun[1], Jin Weon Kim[2], John D. Hunn[1], Jim H. Miller[1], Lance L. Snead[1]
[1]Oak Ridge National Laboratory, P. O. Box 2008, MS-6151, Oak Ridge, TN 37831, USA
[2]Chosun University, Kwang Ju, South Korea

ABSTRACT

The fracture properties of coated SiC layers and their relationship with microstructure and processing condition have been investigated. A crush testing method for hemispherical shell specimens was developed based on the results of finite element (FE) analysis on stress distribution and applied to the evaluation of fracture strength for the chemical vapor deposited (CVD) SiC layers of tri-isotropic (TRISO) fuel particles. A blanket material was used in the crush tests as the FE analysis confirmed that a relatively soft metal (brass foil) inserted between the specimen's convex surface and plunger tip produced a highly uniform stress under the contact area. Nine sets of hemispherical shell specimens were prepared from different versions of coated fuel particles, which include several relevant lots of coated fuel particles: AGR (advanced gas-cooled reactor) fuels, a German reference fuel, and other development fuels available at Oak Ridge National Laboratory. The coatings display variations in grain size and density, which originated from different conditions in coating processes. At least 30 specimens were tested at room temperature for each material and the fracture stress data were analyzed using Weibull statistics. The local fracture stress calculated was converted to the fracture stress for the whole inner surface area of each particle. Mean fracture stress varied with test material in the range of 330 – 650 MPa, however, could not be clearly connected to the microstructural characteristics.

INTRODUCTION

Very high or high temperature gas-cooled reactors have been developed to increase thermal efficiency as well as to apply to hydrogen production, and the carbon/carbide coated particle fuels compacted in graphite matrix have been developed for such reactors [1-4]. In the coated particles the uranium containing fuel kernels are coated with a porous buffer pyrolytic (Py) carbon layer, and then with tri-isotropic (TRISO) coatings (inner PyC, SiC, and outer PyC layers. Among these layers the SiC layer is the most important component for the structural integrity of the particles because it sustains most of the internal pressure caused by fission gas generation [3,4]. Although obtaining mechanical property data, fracture strength in particular, is essential for the design and performance analysis of the fuel particles, no procedure for testing and evaluation has been widely used or standardized. This is not only because the SiC coatings are too small (<1mm in diameter, <50μm in thickness) for any standardized testing setup but also because the size effect in strength is so significant in such small size range that the strength, which usually depends on the loading method, specimen geometry, and coating process etc., is not clearly defined as a material property.

In recent years, it has been attempted to evaluate the fracture strength of SiC coatings using new or modified methodologies such as the internal pressurization method for tubular and hemispherical shell SiC specimens and the crush test (diametrical loading) for both of these specimen types [1,5-7]. In these attempts, the crush testing methods were convenient to perform; however, data analysis was not so simple because of the extremely high stress concentrations in the coating layers. The internal pressurization techniques required tedious procedures or are expected to have limitations for use at high temperatures since they use plastic inserts. To pursue simplicity in testing and more consistency in data, development of a modified crush testing method is attempted in this study.

In the testing, the hemispherical shell SiC specimens are diametrically loaded between a bottom base and a plunger with brass foil below its end. This metal insert was used to produce a highly uniform stress distribution under the loading contact area and then to accurately evaluate the fracture stress from the measured fracture load data by using an analytical solution, which was confirmed by FE analysis. Nine(9) sets of hemispherical shell specimens were prepared from several relevant lots of coated fuel particles. At least 30 specimens were tested at room temperature for each material and the fracture stress data were analyzed using Weibull statistics. Also, the size effect in the fracture strength data and the relationship with microstructure are discussed.

EXPERIMENT AND ANALYSIS

FE analysis and modification of crush testing

Having a reasonable size of uniformly stressed area is necessary because the analytical solution for stress calculation to be used in calculations is obtained for a concentrated load that is uniformly distributed over a defined contact area whose radius is larger than the thickness of the specimen [9]. FE analysis was performed to prove the feasibility of a soft metal insertion as a measure of expanding the uniformly stressed area at the inside wall of the SiC shell specimen during crush testing. This enabled the use of an analytical solution for stress distribution. The main reason for pursuing this crush test method regardless of some complexity in calculation was the future application at high temperature or in an environmental chamber. Figure 1 illustrates the crush test setup, which shows a soft metal film inserted between the hemispherical shell specimen and the plunger.

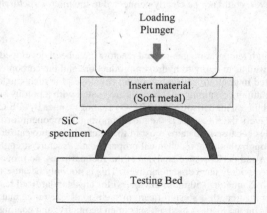

Figure 1. The crush test setup with insert metal.

Figure 2 presents the FE model used in the simulations using the ABAQUS code. Considering the geometrical symmetry of the specimens, an axi-symmetric model employing a four-node bilinear axi-symmetric element with reduced integration (CAX4R) was used. SiC shell specimens were assumed to be fully elastic with Young's modulus of 450 GPa and Poisson ratio of 0.21. The loading plunger was assumed to be a ridged body and for the input data of the insert material (brass), the tensile stress-strain curve was used. A friction coefficient of 0.25 was used for all contact surfaces.

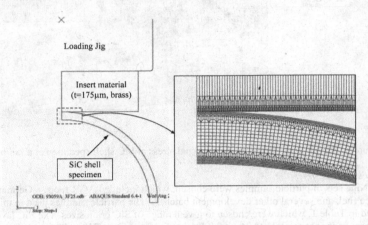

Figure 2. Finite element model used in the analysis

Figure 3 shows the distribution of maximum principal stress, which governs brittle fracture in elastic materials. The highest value of maximum principal stress appeared at the inner surface of the specimen below the loaded area (the contact area with the insert material). For the majority of the inner surface area just below the loading contact area, the stress variation is less than 10% of maximum stress, which justifies the use of the loading contact area for the effective area. Although the uniformly stressed area was slightly less than the loaded area at the outside wall of the specimen, the uniform area was larger than two times the specimen thickness in this example. Based on this FE analysis result, we concluded that the use of a soft metal insert between the plunger and the specimen could produce adequate size and uniformity of the stressed and load-transferring area necessary for the use of the analytical solution.

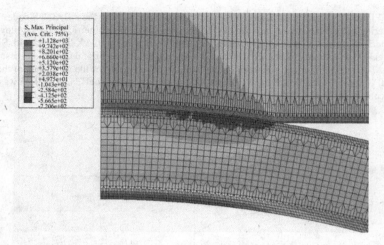

Figure 3. Distribution of maximum principal stress at SiC shell specimen at a failure load

Materials and crush testing

Nine sets of particle samples were chosen, which included AGR fuels, a German reference fuel, and several other development batches. The particle samples used in this study are listed in Table 1, which were chosen to give a range of SiC grain sizes. The SiC layers have thicknesses in the range of 25 to 36 μm and outer diameters from 710 to 890 μm. These coatings also display variations in density, which originated from different conditions in the coating processes. The variation in density and thickness within a given sample set is indicated by the standard deviation in the measured mean given in Table 1.

In the specimen preparation procedure, particles were mounted on aluminum holders using a thermoplastic epoxy (Crystalbond). The particles were ground to near midplane with 9 μm diamond suspension on a cast iron disc (Struer's ALLEGRO). The exposed cross-sections were then polished using 3 μm diamond suspension on a silk disc (Struer's DAC). Kernels that had not already dropped out due to polishing to midplane were removed using adhesive tape. The particles were then removed from the epoxy by reheating the mount and washing off residual epoxy with acetone. Eight hours heating in air at 700°C was sufficient to burn off the carbon layers and yield free-standing SiC hemispherical shells.

A screw-driven tensile machine with a 10 kg capacity load cell was used to load the specimens. For each material, at least 30 specimens were tested at room temperature with a cross head speed of about 0.005 mm/s. The machine was set to detect and display the maximum load, which was used as the fracture load. The diameters of the loading-contact area were measured from the impressions formed at the brass foil, and for each material the average value of the measurements was used in the calculation.

Table 1. Dimensional and density data for hemispherical shell SiC specimens.

No.	Sample ID	Mean Thickness (μm)	Mean Outer Diameter (μm)	Mean Density (g/cc)	Remarks
1	DUN500S-14B	~25	~870	3.185±0.005	Mixed Ar/H SiC deposition at 1340°C, very fine grained and porous
2	DUN500S-6B	~30	~886	3.205±0.001	H only SiC deposition at 1510°C, large grain
3	DUN500S-7B	~35	~862	3.206±0.005	Mixed Ar/H SiC deposition at 1440°C, small grain
4	AGR-06	33.9±1.4	850	3.201±0.002	German reference fuel
5	AGR-10	26.8±0.6	718	3.206±0.002	US HRB-21 reference fuel
6	LEU01-46T	35.3±1.3	759	3.2075±0.0032	AGR-1 Baseline
7	LEU01-49T	35.9±2.1	756	3.2046±0.0010	AGR-1 Variant 3 (Ar-H mixed SiC deposition, finer grain structure at lower deposition temperature)
8	B&W-93059	34.3	~797	3.199	B&W AGR-2 Variant Qualification TRISO
9	B&W-93060	36.8	~813	3.195	B&W AGR-2 Baseline Qualification TRISO

Note: ± values give the measured standard deviation and indicate how much variation was observed for that property.

Calculation of local fracture stress

When a partial spherical shell is diametrically loaded by an external load F concentrated on a small circular area of radius r_0, the stress components in the thin shells, the maximum membrane stress and bending stress, are given by [9]

$$\sigma_{menbrane} = -C_1 \frac{F\sqrt{1-\upsilon^2}}{t^2} \tag{1}$$

$$\sigma_{bending} = -C_2 \frac{F\sqrt{1+\upsilon}}{t^2} \tag{2}$$

where v is Poisson ratio, t is the thickness of shell specimen, R is the outer diameter of the shell, and the coefficients C_1 and C_2 can be given by the fitting equations (produced by the authors):

$$C_1 = 0.2205 - 0.04\mu - 0.0115\mu^2 \tag{3}$$

and

$$C_2 = 1.2044 \exp(-1.2703\mu); \tag{4}$$

$$\mu = r_0[12(1-\upsilon^2)/(R^2t^2)]^{1/4}. \tag{5}$$

Then, the maximum tensile stress which occurs at the inner surface of the shell is given by

$$\sigma_{max} = \sigma_{membrane} + \sigma_{bending} \tag{6}$$

At fracture, this maximum stress becomes the fracture stress, which represents the local loading or the size of sampling. Since the fracture stress is dependent on the loaded volume or area and the size effect is significant in the range of the present specimen size, this local fracture stress is converted to the value for a full size spherical shell.

In the calculations we assumed that all dimensional measurements are the same for each set of specimens, ignoring variations in shape and size of specimen, in loading-contact area measurements, etc. Therefore, the Weibull moduli determined for data sets of fracture load and stress are the same and reflect the effects from such dimensional variations.

Calculation of Weibull parameters

It is well known that ceramic materials are notorious for their widely variable strength among seemingly identical specimens, and for the dependence of strength on the size and surface condition of the specimens [10-13]. Since the maximum tensile stress always occurs at the inner surface of the specimens and this surface has a large density of dimple-like features which supposedly act as stress concentrators, the present statistical analysis assumes that failure initiates at the inner surface of the hemispherical shell specimens [5-7]. Thus the size effect is described based on the effective area. Using the Weibull's two-parameter distribution [12,13], the cumulative probability of failure P is presented by:

$$P = 1 - \exp[-S_E \left(\frac{\sigma_f}{\sigma_0} \right)^m], \tag{7}$$

where σ_f, m, σ_0, and S_E are the fracture stress, the Weibull modulus, the scale parameter, and the effective surface (or the load weighted surface), respectively. The Weibull modulus m, also called the shape parameter, represents the extent of scatter in the fracture strength. The scale parameter σ_0, corresponding to the fracture stress with a failure probability of 63.2%, is closely related to the mean strength of the distribution. The term S_E represents the surface area of a hypothetical specimen subjected to a uniform stress over the whole surface area, which has the same probability of fracture as the test specimen stressed at σ_f. In this analysis the average value of the measured contact areas ($\pi \bar{r}_0^2$) was used for this parameter.

By taking the logarithm twice, Eq. (7) can be rewritten in a linear form:

$$\ln \ln \left(\frac{1}{1-P} \right) = m \cdot \ln \sigma_f + \ln(\frac{S_E}{\sigma_0^m}). \tag{8}$$

The Weibull modulus and scale parameter can be obtained from the slope and intercept terms in Eq. (8), respectively. Since the true value of P_i for each σ_i is not known, a prescribed probability estimator has to be used as the value of P_i. There have been introduced several probability estimators and their merits have been investigated [15,16]. Among those probability estimators, it is shown that the probability estimator of Eq. (7) gives a conservative estimation, and therefore from the engineering point of view it should be the best choice in reliability predictions [15].

$$P_i = \frac{i}{N+1},$$
(9)

where P_i is the probability of failure for the i th-ranked stress datum and N is the sample size.

Size effect and fracture stress for full spherical shell

The effective surface, which includes the effects from specimen geometry, multi-axial stress field, stress gradient, and chosen failure criterion on the reliability of a material, can be used to scale ceramic strengths from one component size to another or from one loading configuration to another [17]. Larger specimens or components are likely to be weaker because of their greater chance to have a larger and more severe flaw. For two specimens having different sizes or loading configurations, the ratio between their mean fracture strengths (or characteristic strengths) can be correlated with the ratio of the effective surface areas [5,6,8]:

$$\sigma_f^F = \left(\frac{S_E^L}{S_E^F}\right)^{1/m} \sigma_f^L = \left(\frac{\pi r_0^2}{4\pi(R-t)^2}\right)^{1/m} \sigma_f^L,$$
(10)

where σ_f^L and σ_f^F are the fracture stresses for partially loaded and full spherical shell specimens, respectively, S_E^L and S_E^F are the effective surfaces. In the evaluation of size effect, the measured radius of indentation impression was used for the radius of effective area (r_0).

RESULTS AND DISCUSSION

Figure 4 displays the Weibull plots of fracture strength data (σ_f^F), and Table 2 summarizes the measured and calculated mean values for key Weibull parameters. In each material the lowest fracture stress was often less than one half of the maximum value. The mean fracture stress varied with test material in the range of 330 – 650 MPa; the values for scale parameter were about 17% higher than the mean fracture stress. Among the test materials B&W-93060 had the lowest fracture stress of about 330 MPa and LEU01-49T the highest strength, about 650 MPa. These SiC strength values are similar to those obtained from the surrogate tubular specimens if size effect is taken into account; for example, the fracture stress obtained from the tubular SiC specimens with ~1 mm diameter and 0.1 mm wall thickness was about 300 MPa [5-7].

The Weibull modulus was in the range of 3.98 – 7.25. Except for LEU01-46T which showed the largest scatter in data or the lowest modulus (3.98), the test materials showed a modulus higher than 5. In determining these modulus values all dimensional variations among specimens were ignored, and therefore, these values are the same for the data sets for fracture load and fracture stress for each material.

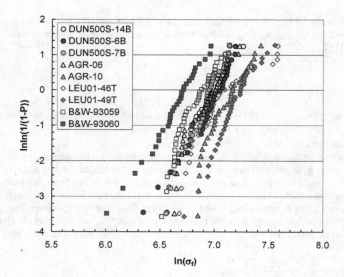

Figure 4. The Weibull plots of fracture strength data for nine archive materials

Table 2. Measured values and Weibull parameters

Material ID	Mean Contact Diameter, mm	Mean Fracture Load, N	Weibull Modulus	Mean Fracture Stress (Scaled to Whole Shell), MPa
DUN500S-14B	0.118	2.59	6.61	449.8
DUN500S-6B	0.125	3.60	5.49	409.6
DUN500S-7B	0.142	5.32	7.25	514.7
AGR-06	0.147	5.84	6.22	475.4
AGR-10	0.113	4.20	6.40	570.7
LEU01-46T	0.153	11.36	3.98	399.1
LEU01-49T	0.141	8.64	6.35	646.5
B&W-93059	0.151	6.47	6.58	463.9
B&W-93060	0.167	6.97	5.15	329.9

SEM images show the relative grain size for the different samples. DUN500S-14B exhibited the smallest grain size. LEU01-49T and B&W-93060 ranked next in size, with B&W-93059, DUN500-7B and AGR-06 (German) just slightly larger in average grain size. LEU01-46T had noticeably larger grains in the outer portion, but would still be ranked as having an acceptably fine grain structure. DUN500-6B and AGR-10 (HRB-21) possessed relatively large grains, some extended up to half the total layer thickness. The finer grain structures also appeared to be more porous but did not result in lower strength in a consistent manner. Figure 5 shows examples of the SEM images.

Figure 5. Microstructures of SiC layers: (a) AGR-06 German Reference, (b) Fuel LEU01-46T AGR-1 Baseline.

It has been believed that the microstructural and geometrical parameters, such as grain size, surface roughness, porosity, and irregularities in thickness and diameter, determine the strength characteristics of SiC coatings. Although more focused and detailed investigation is needed for a final conclusion, the following paragraphs discuss the influences from those parameters:

Although it is expected that a larger grain size induces lower fracture stresses because of higher stress concentrations at grain boundaries, the influence of grain size on the fracture stress is not clear in this study. As described in Table 1, it is found that among the large grain materials are DUN500S-6B and ARG-10. However, their fracture stresses are not among the lowest although the fracture stress for DUN500S-6B is the third lowest one (Table 2). This indicates that the grain size is not a dominant factor determining the fracture process of a shell specimen.

The roughness of the inner surface might be the controlling factor for the initiation of fracture since the maximum stress always occurs at the inner surface where the dimple-like structure provides plenty of crack initiation sites [5]. Note that the scaling using the effective area, Eq. (10), is based on the notion that the fracture initiates at the inner surface of the shell specimens. SEM images [8] show high porosity near the inner surface of the coatings and SiC infiltrates into the inner PyC layer. The degree of roughness, however, is not discernable for those coatings. Therefore, seemingly the most influential microstructural property cannot explain the difference of the fracture stress data among the materials.

The variances in the mean diameter and thickness, along with the shape irregularities in a specimen, should affect the strength result significantly. As listed in Table 1, however, we have very limited data for those parameters. Comparing the data for LEU01-46T and 49T, the specimen set with a larger standard deviation in thickness did not produce a larger scatter (or smaller Weibull modulus) in fracture stress data. It is believed that the deviation in the thickness is too small to produce a discernable effect in the strength results or its effect is compensated by other effects.

Finally, the size effect was significant: the scaling factor between the local fracture stress and the fracture stress for full size shell in the range 1.9 – 3.1. The fracture stress in the present specimen size range is considered not as a material property but as a property for a material plus a specific structure. Therefore, in the application of the data the effective area for the specific phenomenon should be evaluated and the fracture stress data provided be scaled up or down using the Eq. (10).

CONCLUSIONS

[1] The crush test and evaluation method using a deformable metallic foil at the specimen-plunger contact was successfully applied to the evaluation of fracture stress for the hemispherical shell SiC specimens.

[2] The test method was found to be self consistent and showed reasonably well controlled scatter between specimens in a given sample. The average SiC strength for shells typical of those found in TRISO particles was found to be in the range of 330-650 MPa. The statistical characteristics of fracture stress are consistent with the earlier data for spherical shell or tubular specimens.

[3] The statistical characteristics of fracture stress were not well explained by the varied microstructural characteristics of the materials studied. This might be because the degree of roughness at the inner surface, the main fracture controlling factor, was not discernable for different coatings while the fracture stress data evaluated were believed to result from the competition of effects from multiple parameters.

[4] The size effect should be always considered in application of the data. The scale factor between the current shell specimen size and the stressed size in the new application can be calculated by Eq. (10) or other known equations.

ACKNOWLEDGEMENTS

This research was sponsored by the Office of Nuclear Energy Science and Technology, US Department of Energy under contract DE-AC05-00OR22725 with UT-Battelle, LLC. The authors thank Mr. I. Dunbar for preparing hemispherical shell specimens.

REFERENCES

[1] K.E. Gilchrist and J.E. Brocklehurst, "A Technique for Measuring the Strength of High Temperature Reactor Fuel Particle Coatings," *J. Nucl. Mater.*, *43*, 347-350 (1972).

[2] P.L. Allen, L.H. Ford, and J.V. Shennan, "Nuclear Fuel Coated Particle development in the Reactor Fuel Element Labortories of the U.K. Atomic Energy Authority," *Nucl. Tech.*, 35, 246-353 (1977).

[3] G.K. Miller, D.A. Petti, D.J. Varacalle, and J.T. Maki, "Statistical Approach and Benchmarking for Modeling of Multi-dimensional Behavior in TRISO-coated Fuel Particles," *J. Nucl. Mater.*, **317**, 69-82 (2003).

[4] D.A. Petti, J. Buongiorno, J.T. Maki, R.R. Hobbins, and G.K. Miller, "Key Differences in the Fabrication, Irradiation and High Temperature Accident Testing US and German TRISO-coated Particle Fuel, and Their Implications on Fuel Performances," *Nucl. Eng. Des.*, **222**, 281-297 (2003).

[5] S.G. Hong, T.S. Byun, R.A. Lowden, L.L. Snead, Y. Katoh, "Evaluation of the Fracture Strength for SiC Layers in the TRISO-coated Fuel Particle," *J. of the Amer. Ceramics. Soc.*, **90** 184-191 (2007).

[6] T.S. Byun, E. Lara-Curzio, L. L. Snead, Y. Katoh, "Miniaturized Fracture Stress Tests for Thin-Walled Tubular SiC Specimens," *J. of Nucl. Mater.*, **367-370**, 653-658 (2007).

[7] L. L. Snead, T. Nozawa, Y. Katoh, T. S. Byun, S. Kondo, D. A. Petti, "Handbook of SiC Properties for Fuel Performance Modeling," *J. of Nucl. Mater.*, 371, **329-377** (2007).

[8] T.S. Byun, J.W. Kim, I. Dunbar, J. D. Hunn, "Fracture Stress Data for SiC Layers in TRISO-Coated Fuel Particles" Oak Ridge National Laboratory Research Report, ORNL/TM-2008/167, September (2008).

[9] R.J. Roark, W.C. Young, "Formulas for Stress and Strain," McGraw-Hill Book Co., Fifth Eds, New York, 1974.

[10] M. Jadaan, D.L. Shelleman, J.C. Conway, Jr., J.J. Mecholsky, Jr., and R.E. Tressler, "Prediction of the Strength of Ceramic Tubular Components: Part I - Analysis," *JTEVA*, **19**, 181-191 (1991).

[11] D.L. Shelleman, O.M. Jadaan, D.P. Butt, R.E. Tressler, J.R. Hellman, and J.J. Mecholsky, Jr., "High Temperature Tube Burst Test Apparatus," *JTEVA*, **20**, 275-284 (1992).

[12] ASTM Standard, "C1239-00 Standard Practice for Reporting Uniaxial Strength Data and Estimating Weibull Distribution Parameters for Advanced Ceramics," American Society for Testing and Materials, Philadelphia, PA (2003).

[13] M.A. Madjoubi, C. Bousbaa, M. Hamidouche, and N. Bouaouadja, "Weibull Statistical Analysis of the Mechanical Strength of a Glass Eroded by Sand Blasting," *J. Eur. Ceram. Soc.*, **19**, 2957-2962 (1999).

[14] T. Lin, A.G. Evans, and R.O. Ritchie, "A Statistical Model of Brittle Fracture by Transgranular Cleavage," *J Mech. Phys. Solids*, **21**, 263-277 (1986).

[15] B. Bergman, "On the Estimation of the Weibull Modulus," *J. Mater. Sci. Lett.*, **3**, 689-692 (1984).

[16] A. Khalili and K. Kromp, "Statistical Properties of Weibull Estimators," *J. Mater. Sci.*, **26**, 6741-6752 (1991).

[17] D.G.S. Davies, "The Statistical Approach to Engineering Design in Ceramics," *Proc. Br. Ceram. Soc.*, **22**, 429-452 (1973).

OPTIMIZATION OF FRACTURE STRENGTH TESTS FOR THE SIC LAYER OF COATED FUEL PARTICLES BY FINITE ELEMENT ANALYSIS

Jin Weon Kim[1], Thak Sang Byun[2], and Yutai Katoh[2]
[1] Department of Nuclear Engineering, Chosun University, Gwangju, Korea
[2] Material Science and Technology Division, Oak Ridge National Lab, Oak Ridge, TN 37831, USA

ABSTRACT

Extended finite element (FE) analysis has been performed to optimize the fracture strength tests for SiC coating layer of tri-isotropic (TRISO) fuel particles. Since the SiC layer is a thin spherical shell that sustains most of the internal pressure from fission gas release and other internal expansions, obtaining accurate mechanical property data has been required for design optimization and performance assessment of the fuel. A crush testing technique for hemispherical shell SiC specimens has been newly developed, in which a metallic foil is inserted between the SiC specimen and the loading plunger to produce adequate stress distribution in the specimen. The present FE analysis aimed at finding proper material and thickness for the metallic insert, and detailed parametric studies were performed for insert material and thickness. The results could suggest optimum loading configurations that can provide sufficient uniformly stressed area, on which the reliability of fracture strength data is dependent. Fracture tests were conducted for model specimens, following the suggested details of loading configuration. The results demonstrated that the suggested loading configuration produced reliable fracture strength data for the SiC coating layer of TRISO fuel particles.

INTRODUCTION

A tri-isotropic (TRISO) carbon/carbide-coated fuel particle is typically composed of an oxide (or carbide) fuel kernel and four layers of coatings [1,2]. The SiC layer among the TRISO coating layers is the most important component since it not only provides the structural integrity for whole fuel particle but also retains fission products [3-6]. Therefore, the production of reliable fracture strength data for the thin SiC layer has been required to use in the design and performance analyses for fuel particles. It is known that the fracture strength data show a strong size effect in the small size range (<1 mm in diameter and <50 μm in thickness), and thus non-standard testing and evaluation methods have been attempted [5-8].

One of the methods is a crush testing technique on hemispherical shell specimens [5]. In the crush test, the SiC shell specimen is directly crushed under a compressive loading on the top of the hemispherical shell specimen using a flat-ended loading plunger, and the fracture strength is determined from a load versus maximum principal stress relationship given by elastic finite element (FE) analysis. In this method, however, the calculated stress distribution in the hemispherical shell is strongly dependent on the contact condition between the loading plunger and the SiC specimen. In general, the contact area is determined by the elastic deformation of the shell specimen and the roughness of outer surface of the specimen. As a result, the load versus maximum principal stress relation is normally unknown and is dependent on the conditions of contact area. It is believed that this causes significant errors in any simulation or theoretical calculation to obtain the fracture strength data. Lately, a new crush testing technique for hemispherical shell SiC specimens has been developed at Oak Ridge National Laboratory, as illustrated in Fig. 1. This technique uses a metallic foil inserted between the specimen and the loading plunger to produce an adequate stress distribution in the specimen.

This FE study is part of the attempt to develop a new testing and evaluation method, and aimed at searching for an optimum material and thickness for the metallic foil insert. This insert is plastically deformed during compression and the contact area can be determined by the plastic deformation in the insert, instead of the flattened area on the outer surface of SiC specimens. It is expected that the errors in calculations caused by uncertain contact area is eliminated because the contact area can be measured from the impression in the insert. It is also expected that the increase in loading area by the deformation of insert metal is able to expand the uniformly stressed area at the inner surface under the loading contact. The results of this FE analysis will prove the effect of insert metal on the stress distribution and provide optimum conditions for using a metallic insert in the crush testing. As a confirmation test, the modified test method is applied to a model shell SiC coating.

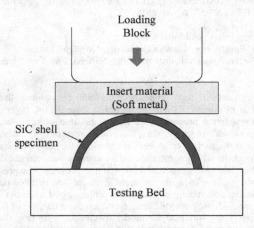

Fig. 1. Setup for modified crush test for hemispherical shell SiC specimens

PARAMETRIC ANALYSIS

Finite Element Model for Crush Test
    In order to find an optimum loading configuration and insert material conditions, a series of parametric FE analyses were conducted using a general purpose FE program, ABAQUS [9]. Figure 2 presents the FE model used in the analysis. Considering the geometrical symmetry of the specimens, an axi-symmetric model of four-node element with reduced integration (CAX4R) was used in the FE modeling. Fine meshes were employed for all portions of interest, i.e., the load transferring contact surfaces and the inner surface of SiC shell to obtain the accurate stress distribution. The thickness and outer radius of the SiC shells are 49μm and 423μm, respectively. It was assumed that the deformation of a hemispherical shell SiC specimen was fully elastic and isotropic, and thus its elastic behavior is described by a Young's modulus of 450 MPa and a Poisson's ration of 0.21. The loading plunger (cylindrical steel rod) was also assumed to be an elastic body with a Young's modulus of 210 MPa and a Poisson's ration of 0.29.

    Five different insert metals were simulated in the parametric analysis: annealed Cu (Cu_Annl.), modified Cu (Cu_mod.), annealed SS316 (SS316 Annl.), cold worked SS316 (SS316_CW), and brass. Figure 3 presents the true stress-true strain curves used in the present parametric FE analysis. Stress distribution was simulated for various metallic materials with different strengths and strain hardening behaviors. Also, five different thicknesses of insert metal were simulated in the parametric analysis: $t_I =$ 50, 100, 150, 200, and 250μm. Friction coefficient was assumed to be 0.25 for all contact surfaces: between the SiC shell outer surface and the end surface of the insert and between the insert metal and the loading plunger.

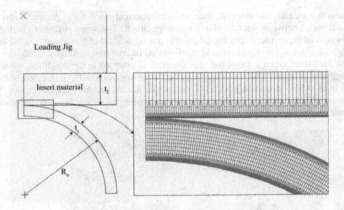

Fig. 2 Finite element analysis model

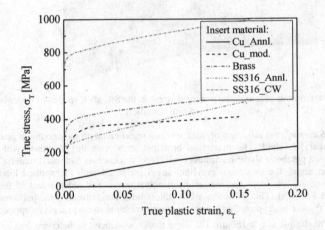

Fig. 3. True stress-true strain curves used in the parametric finite element analyses

Results of Parametric FE Analysis

Using the results of FE analysis, the stress distribution, load versus maximum principal stress relationship, and effective surface area were calculated and investigated. Figure 4 presents the contour of principal stress in the hemispherical SiC shell specimen under compressive load. As expected, it shows that the principal stress was concentrated at the center region of inner surface of the specimen. Figure 5 compares the variations of principle stress distribution at the inner surface of specimen. Here, the stress was normalized with respect to the maximum value. Figure 5(a) clearly reveals that the uniformly stressed area at the inner surface of hemispherical shell SiC specimen was enlarged by inserting a soft metal between loading plunger and specimen. The stress distribution slightly changed with the thickness of insert metal, but the effect was not significant at $t_l > 50\mu m$. On the other hand, as shown in Fig. 5(b), the stress distribution at the inner surface of hemispherical shell was considerably changed with load level when a metal foil was inserted, in contrast to the condition without insert that shows negligible variation. The uniformly stressed area was gradually expanded with increase in compressive load. Above a certain load level, moreover, the maximum stressed location moved away from the center of hemispherical shell surface. In particular, the deviation was significant when the very soft metal, such as Cu_Annl., was used as insert foil.

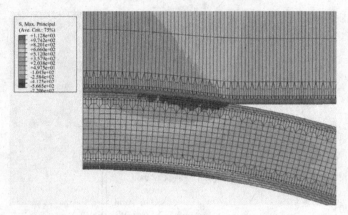

Fig. 4. Contour of maximum principal stress in the SiC shell specimen under compressive loading condition

Figure 6 displays the influence of thickness and insert metal on the maximum principal stress in the hemispherical SiC shell. The maximum principal stress (tensile) increased with applied load (compressive) in a parabolic shape regardless of different thicknesses and insert materials. Even in the case without an insert, the maximum principal stress increased with the applied load in a slightly nonlinear manner. At low loads the maximum principal stress was independent of the thickness of insert shown in Fig. 6(a). The influences of thickness on the maximum principal stress appeared at higher loads; a thicker insert provided lower maximum principal stress at a given applied load, but the difference was negligible at $t_l \geq 150\mu m$. The same thickness-dependent behavior was observed for the other insert metals. Comparing insert materials, it is seen that the magnitude of maximum principal stress is almost proportional to the strength of insert material. The higher strength insert shows higher maximum principal stress at a given load level, which results from a higher stress concentration. With a soft insert the maximum principal stress increased with increasing compressive load, and then was nearly saturated at a certain level. With a high strength insert, however, it continuously increased with load to higher level.

The effective surface area is also an important parameter in the statistical evaluation of fracture strength, in particular, when the fracture initiation mechanism in specimens is governed by surface condition. The SiC coating in fuel particles, whose inner surface morphology is determined by the surface condition of pyrocarbon substrate, is a typical case to describe its property using the effective surface area. Here, the effective surface area represents the surface area of a hypothetical specimen subjected to a uniform stress of $\sigma_{pl,max}$ over the whole specimen area [5].

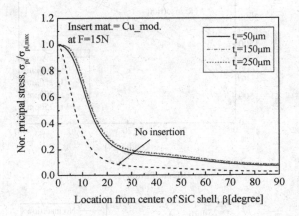

(a) Effect of thickness

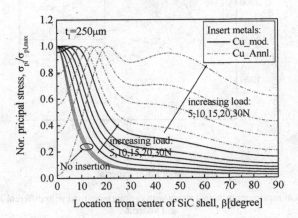

(b) Effect of insert material

Fig. 5. Variations of principal stress distribution at the inner surface of SiC shell by insertion of metallic foil

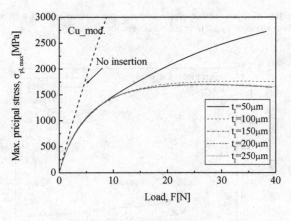

(a) Effect of thickness

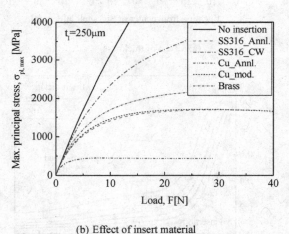

(b) Effect of insert material

Fig. 6. Variations of maximum principal stress with compressive load with different insert thicknesses and materials

In this work, the principal stress distribution was obtained from FE analysis results, so that the effective surface area was calculated by numerical integration using the following equation [5]:

$$S_E = \sum_i \left[\left(\frac{\sigma_{1i}}{\sigma_{max,pl}} \right)^m + \left(\frac{\sigma_{2i}}{\sigma_{max,pl}} \right)^m \right] S_i , \qquad (1)$$

where σ_{1i}, σ_{2i}, and S_i are the two principal tensile stress components at inner surface and the surface area in the i-th element in the mesh of a specimen, respectively. In Eq. (1), m is the Weibull modulus that presents the scatter in the fracture strength and is given by a Weibull plot of fracture load or strength data. In the parametric analyses, the value of m was assumed to be 5.8. Figure 7 shows the load versus effective surface area relationship for different thicknesses and materials. For all cases, the effective surface increased with increasing compressive load. As shown in Fig. 7(a), the effective surface area increased significantly by employing the metallic insert. Its overall dependence on the thickness and material inversely resembled those of the maximum principal stress. At a lower load level the effective surface was independent of the thickness of insert. At a higher load level, however, the effective surface was affected by the thickness of insert, but the thickness dependence was negligible at $t_1 \geq 150\mu m$ for all materials. Also, the softer insert metal has the larger effective surface as shown in Fig. 7(b). When the annealed Cu, a very soft metal, was used, the effective surface area increased rapidly with increasing compressive load. This should be associated with the significant change in the stress distribution during compressive load. As the load increased, the maximum stress location moved away from the center, which resulted in larger stressed area.

From the results of parametric FE analysis, it is confirmed that the uniformly stressed area in the inner surface of hemispherical shell SiC specimen is enlarged by using a metallic insert. It is seen that the thickness of insert has no effect on the maximum principle stress and effective surface area when the insert is thicker than 150μm. The maximum principal stress increased with increasing in the strength of insert, but the effective surface area decreased. The expansion of effective surface is a positive effect on the fracture strength test for ceramic material. If the insert metal is too soft, however, the variation in the stress distribution with load level becomes significant and thus it is expected the location to start fracture is not consistent in a set of test data. Therefore, it can be concluded that the metal with intermediate strength and thicker than 150μm is proper as an insert for crush test using hemispherical SiC shell specimen.

APPLICATION

Crush Test for SiC Shell Specimen

In order to verify the improved loading configuration, a series of crush tests were conducted using model SiC specimens. Modified Cu foil with a thickness of 250μm and brass foil with a thickness of 175μm were selected as inserts. These have intermediate strengths and thicknesses more than 150μm, which are recommended in the parametric analysis above. True stress-true strain curves of those metals are presented in Fig. 3. The thickness and outer radius of hemispherical shell SiC specimens tested are 51μm and 423μm, respectively. Detailed descriptions on the manufacturing process and preparation of hemispherical shell specimens can be found in Ref.[5,10].

In testing, about 30 specimens (per each condition) were tested at room temperature for the three test configurations including the test without insert. The top surface of a specimen was compressed by loading plunger with or without insert. A screw-driven universal testing machine with a 10 kgf capacity load cell was used to load the specimen at a cross head speed of about 0.005mm/sec. The machine was set to catch and display the maximum load, which was regarded as the fracture load (F_f). A summarized test data are given in Table I. The mean fracture loads were almost the same for the cases using the modified Cu or brass insert, and they were higher than those without insert metal.

Table I. Summary of Fracture Load Measurements

Test condition	Number of samples	Averaged fracture load, $\overline{F}_f$ [N]	Standard deviation of fracture load [N]
No insertion	29	8.9	1.49
Cu_mod. (t_l=250 µm)	31	11.1	2.01
Brass (t_l=175µm)	27	11.2	2.37

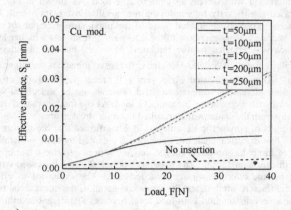

(a) Effect of thickness

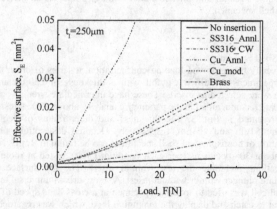

(b) Effect of insert material

Fig. 7. Variations of effective surface with compressive load for different thicknesses and insert materials

Data Analysis and Results

As shown in the results of parametric analysis, the stress in the hemispherical shell SiC specimen under compressive load is concentrated around the center of inner surface, and the stress level varied non-linearly with the strength of insert material and applied load. Stress distribution in specimen was calculated using FE analysis and the load versus maximum principal stress relationships were obtained for all the test configurations. Then, the local fracture strength corresponding to a measured fracture load was calculated using the relationship obtained by FE calculation. Data treatment and expression were carried out based on the two parameter Weibull statistics [11,12] and on the definition of failure probability estimator (P(i) = i/(1+N), where i and N are the rank and total number of data.) [13]. The effective surface area is used to scale ceramic strength from one component size to another or from one loading configuration to another [5,6,10,14]. For different specimen sizes or loading configurations, therefore, the ratio between their mean fracture strengths can be correlated with the ratio of the effective surfaces shown in Eq. (2):

$$\frac{\sigma^A}{\sigma^B} = \left(\frac{S_E^A}{S_E^B} \right)^{-1/m} \tag{2}$$

where σ^A and σ^B are the mean strengths of type A and B specimens or loading configurations, which may have different sizes and stress distributions, and S_E^A and S_E^B are the corresponding effective surfaces. Thus, the fracture strength of a spherical shell SiC specimen under uniform loading by internal pressure can be estimated using Eq. (2) from the local fracture strength and the ratio of effective surfaces.

Local fracture strengths and effective surfaces corresponding to the measured fracture loads were obtained for all test conditions. The local fracture strengths were simply calculated from the load versus maximum principal stress relationship for each test condition (see Fig. 6). The effective surface area is a function of Weibull modulus and normalized-stress weighted area [11-14]. The local fracture strength and effective surface area became lower and larger, respectively, when a metallic insert was used. Table II also presents that the mean local fracture strength with modified Cu insert was about a half of that without insert. This is explained by the difference in the stressed areas with different loading configurations.

In Table II, the mean values of fracture strength, effective surface area, scale parameter are listed for the three test configurations per dataset used to obtain Weibull modulus. The Weibull moduli from fracture load data were in the range of 5.13 - 6.48 for the three test configurations (Case I), displaying only minor differences among the moduli. If the moduli are obtained from the local fracture strength datasets, however, the moduli are significantly different for different test configurations. The moduli with modified Cu or brass insert were about 2 -3 times higher than that without insert. This configuration dependent relationship among Weibull moduli is associated with the non-liner relationship between the local fracture stress and the measure fracture load. As indicated in Fig. 6, a data range selected from the load axis is always larger than the corresponding data range in the local fracture strength axis. This indicates that the higher Weibull modulus is obtained with insert in localized loading.

As Eq. (2) indicates, the fracture strength and scale parameter using higher m-values were higher than those obtained using the Weibull modulus of fracture load data. In the cases II and III, while the difference between the Weibull and strength parameters are insignificant, the fracture strengths and scale parameters with insert showed still lower than those without insert. In addition, the Weibull moduli could be evaluated from the fracture strength data, which are for the uniform loading on the whole spherical shell and are converted from the local fracture strength data using Eq. (2). It should be understood here that the fracture strength and m-value are related each other, and therefore, the Weibull moduli for Cases II and III were obtained by iterative calculation, while those for fracture load data are experimentally measured. In the case III, the moduli with modified Cu and brass inserts were about 1.5 - 2 times higher than those without insert. The fracture strength and scale parameter using the listed Weibull moduli were slightly lower than those evaluated using Weibull moduli of the local fracture strength data in Case II. The fracture strength and scale parameter with insert were lower about 10% than those without insert. Overall, the fracture strength and scale parameter with insert were always lower than those without insert. Also, the fracture strength and scale parameter were weakly dependent on the insert material if the insert materials are in a group of similar strengths.

Equation (2) indicates that the fracture strengths, if scaled to the same effective area, should be the same for the cases with and without insert. In the present result, however, the fracture strength and scale parameter were always higher without insert. This is because of the overestimation of local fracture strength without insert metal, which is induced by disregard of initial contact area in the FE analysis that is unknown. Without using an insert, the real contact area can be hardly evaluated in FE analysis because the crush of surface structure cannot be easily understood. It is again highlighted that the errors in the fracture strength of SiC shell specimen induced by uncertain initial contact area can be eliminated by using deformable insert between loading plunger and hemispherical shell specimen.

Table II. Summary of Fracture Strength Data for SiC Shell Specimen Obtained by Modified Crush Test

Parameter/test configuration (insert)		Cu_mod.	Brass	No insert
Mean local fracture strength, $\bar{\sigma}_{L,f}$ [MPa]		1484.6	1721.4	2822.3
Case I: Based on fracture load data	Weibull modulus, m	5.98	5.13	6.48
	Mean effective surface area, $\bar{S}_E$ [mm^2]	0.00781	0.00715	0.00179
	Mean fracture strength, $\bar{\sigma}_{CV,f}$ [MPa]	600.5	598.0	976.8
	Mean scale parameter, $\bar{\sigma}_o$ [MPa]	721.1	738.3	1146.7
Case II: Based on local fracture strength data*	Weibull modulus, m	18.28	13.49	7.61
	Mean effective surface area, $\bar{S}_E$ [mm^2]	0.00313	0.00309	0.00150
	Mean fracture strength, $\bar{\sigma}_{CV,f}$ [MPa]	1049.9	1090.9	1117.1
	Mean scale parameter, $\bar{\sigma}_o$ [MPa]	1111.6	1177.3	1278.2
Case III: Based on fracture strength data**	Weibull modulus, m	14.34	10.58	7.3
	Mean effective surface area, $\bar{S}_E$ [mm^2]	0.00383	0.00381	0.00157
	Mean fracture strength, $\bar{\sigma}_{CV,f}$ [MPa]	968.5	978.1	1081.1
	Mean scale parameter, $\bar{\sigma}_o$ [MPa]	1042.0	1078.8	1244.7

*The local maximum stress at fracture in the concentrated stress field; **The maximum stress at fracture for the whole spherical shell.

CONCLUSIONS

In order to optimize the suggested test method using insert between loading plunger and specimen, parametric FE analyses were performed. Tests on model hemispherical shell SiC specimens were also carried out using the inserts selected based on the FE simulation results. The following conclusions were drawn from the FE analysis and mechanical test results:
(1) The uniformly stressed area at the inner surface of hemispherical shell could be expanded by using metallic insert between the loading plunger and the specimen.
(2) The maximum principle stress and effective surface area were independent of insert thickness when

using an insert thicker than 150μm.
(3) The maximum principal stress decreased but the effective surface area increased as the strength of insert decreased. With a very soft insert, however, the dependence of stress distribution on the applied load becomes significantly different from those of the medium or high strength insert.
(4) Consistent fracture strength data can be produced for the thin SiC coating by choosing an appropriate insert material and its thickness. A medium strength metallic foil (brass, nickel, etc.) thicker than 150 μm is recommended to use for room temperature testing.

REFERENCES
[1] L.L. Snead, T. Nozawa, Y. Katoh, T.S. Byun, S. Kondo, D.A. Petti, Handbook of SiC Properties for Fuel Performance Modeling, *J. Nucl. Mater.*, **371**, (2007) 329.
[2] L. Tan, T.R. Allen, J.D. Hunn, J.H. Miller, EBSD for Microstructure and Property Characterization of the SiC-coating in TRISO Fuel Particles, *J. Nucl. Mater.*, **372**, (2008) 400.
[3] G.K. Miller, D.A. Petti, J.T. Maki, and D.L. Knudson, An Evaluation of the Effects of SiC Layer Thinning on Failure of TRISO-coated Fuel Particles, *J. Nucl. Mater.*, **355**, (2006) 150.
[4] G.K. Miller, D.A. Petti, J.T. Maki, and D.L. Knudson, Update Solution for Stresses and Displacements in TRISO-coated Fuel Particles, *J. Nucl. Mater.*, **374**, (2008) 129.
[5] S.G. Hong, T.S. Byun, R.A. Lowden, L.L. Snead, Y. Katoh, Evaluation of the Fracture Strength for Silicon Carbide Layers in the Tri-Isotropic-Coated Fuel Particle, *J. Am. Ceram. Soc.*, **90** (2007) 184.
[6] T.S. Byun, E. Lara-Curzio, R.A. Lowden, L.L. Snead, Y. Katoh, Miniaturized Fracture Stress Tests for Thin-Walled Tubular SiC Specimens, *J. Nucl. Mater.*, **367-370**, (2007) 653.
[7] O.M. Jadaan, D.L. Shelleman, J.C. Conway Jr., J.J. Mecholsky Jr., R.E. Tressler, Prediction of the Strength of Ceramic Tubular Components: Part I – Analysis, *JTEVA*, **19** (1991) 181.
[8] D.L. Shelleman, O.M. Jadaan, D.P. Butt, R.E. Tressler, J.R. Hellman, J.J. Mecholsky Jr., High Temperature Tube Burst Test Apparatus, *JTEVA*, **20** (1992) 275.
[9] ABAQUS Analysis User's Manual, Version 8.7, 2008.
[10] T.S Byun, J.W. Kim, I. Dunbar, J.D. Hunn, Fracture Stress Data for SiC Layers in the TRISO-Coated Fuel Particles, *ORNL/TM-2008/167*, Sept. 2008.
[11] ASTM Standard, C1239-00 Standard Practice for Reporting Uniaxial Strength Data and Estimating Weibull Distribution Parameters for Advanced Ceramics, *American Society for Testing and Materials*, Philadelphia, AP (2003).
[12] M.A. Madjoubi, C. Bousbaa, M. Hamidouche, N. Bouaouadja, Weibull Statistical Analysis of the Mechanical Strength of a Glass Eroded by Sand Blasting, *J. Eur. Ceram. Soc.*, **19** (1999) 2957.
[13] B. Bergman, "On the Estimation of the Weibull Modulus, *J. Mater. Sci. Lett.*, **3** (1984) 689.
[14] D.G.S. Davies, The Statistical Approach to Engineering Design in Ceramics, *Proc. Br. Ceram. Soc.*, **22** (1973) 429.

LASER MELTING OF SPARK PLASMA SINTERED ZIRCONIUM CARBIDE:
THERMOPHYSICAL PROPERTIES OF A GENERATION IV VERY HIGH TEMPERATURE
REACTOR MATERIAL

Heather F. Jackson[1], Doni J. Daniel[1], William J. Clegg[2], Mike J. Reece[3], Fawad Inam[3], Dario Manara[4], Carlo Perinetti Casoni[4], Franck De Bruycker[4], Konstantinos Boboridis[4], William E. Lee[1]

[1]Department of Materials, Imperial College London, London, United Kingdom
[2]Department of Materials Science and Metallurgy, University of Cambridge, Cambridge, United Kingdom
[3]Nanoforce Technology Limited, London, United Kingdom
[4]Institute for Transuranium Elements, European Commission, JRC, Karlsruhe, Germany

ABSTRACT
    Commercial $ZrC_{0.96}$ powder (ABCR, Karlsruhe, Germany) was densified by spark plasma sintering to greater than 96% relative density at temperatures of 1900-2180°C, applied pressures of 40-100 MPa, and soak time of 6-30 min. Effects of process parameters on microstructure were assessed by ceramography. High temperature (>2000°C) was more instrumental in full densification than was high pressure, and excessive ramp rate resulted in high residual porosity. Grain coarsening was promoted by prolonging the isothermal dwell.
    Laser heating was used to melt sintered ceramics, as part of a novel thermal analysis technique for probing extremely high temperature phase transformations. Temperatures well in excess of the expected melting temperature of ZrC and up to 4000 K were achieved. The feasibility of the technique for detecting melting transitions in zirconium carbide was demonstrated, and solidus and liquidus temperatures within 50-80 K of predicted values were measured. Post-melting analysis of laser-melted specimens revealed dendritic microstructure and composition consistent with single phase ZrC.

INTRODUCTION
    The pursuit of ever more energy-efficient, sustainable, and passively safe reactor designs for next-generation nuclear power has led to concepts such as the Very High Temperature Gas-Cooled Reactor (VHTR)[1] whose core, based on the ceramic composite tri-structural isotropic coated particle (TRISO) fuel, is expected to operate above 1250℃ and to tolerate excursions to 1600℃ under accident conditions.
    Transition metal carbides[2,3] combine the mechanical strength and high temperature degradation resistance of ceramics with thermal conductivities and close-packed crystallographic structure more like metals. These properties are interesting for typical hardness and wear-resistance technological applications, but in the case of zirconium carbide (ZrC), low thermal neutron capture cross-section opens the possibility of high temperature nuclear applications.
    In TRISO fuel, ZrC would be utilised as a dense barrier against fission product diffusion and gas release, which at the same time maximises transfer of fission heat and neutrons. Increasing operating temperatures of VHTR beyond those encountered in existing high-temperature prototypes will increase the complexity and variety of potential materials interactions among the fuel components.
    Before ZrC-based TRISO is viable for this application, a more complete understanding of the behaviour of ZrC and TRISO at the temperatures and burnups involved must be developed. For instance, the well-documented tolerance of transition metal carbides for structural vacancies (for ZrC_{1-x}, x ~ 0 to 0.5) is believed to impart stability under irradiation[4,5]. An understanding of the variation of thermophysical and mechanical properties as a function of non-stoichiometry becomes necessary in order to develop a reliable nuclear fuel and wasteform.

The objective of the present study is to demonstrate the feasibility of a novel thermal analysis technique, via laser heating, for accurate measurement of thermophysical properties of ZrC at extremely high temperatures. This technique is applied to the measurement of melting temperatures of ZrC_{1-x}, which yields information on high-temperature degradation as well as insight into the underlying physics of bonding. The study started by reproducing established data as tested on commercially available ZrC and ultimately aims to contribute to areas of the Zr-C phase diagram where experimental data are weak or missing.

The most recent Zr-C phase diagram is based on the assessment by Fernandez-Guillermet[6], depicted in Figure 1 as plotted using the Thermo-Calc code[7]. The picture has remained largely unchanged since the 1960s experimental programmes undertaken by Sara[8,9] and Rudy *et al*[10]. Within the phase diagram a single intermediate compound exists, the monocarbide, with a wide phase field exhibiting only carbon-deficient non-stoichiometric compositions and melting congruently at approximately $ZrC_{0.85}$ (46 at% C). The ZrC_x-C eutectic temperature is well-established, though few liquidus measurements are reported for this two-phase region at compositions much higher than 50 at% C. The $Zr-ZrC_x$ eutectic is very close to the melting temperature of pure bcc Zr, leading to the assumption of nearly zero solid solubility of C in Zr. Liquidus measurements in this two-phase region are absent below approximately 37.5 at% C, the lower phase boundary for ZrC_x.

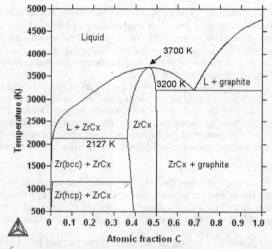

Figure 1. The Zr-C phase diagram, assessed by Fernandez-Guillermet[6] and plotted for this work using the Thermo-Calc code[7].

High-temperature phase transitions in the Zr-C system have been studied by traditional thermal analysis techniques[10] (e.g. differential thermal analysis, DTA), observation of melting (e.g. sample collapse or blackbody hole filling with liquid) during induction[8,9] or resistance[10,11] heating, and observation of incipient melting based on post-quenching ceramography[9,10]. All techniques have associated shortcomings: DTA as a non-isothermal technique is non-equilibrium in nature, direct furnace heating destroys the sample upon melting and precludes study of cooling transitions, while ceramography depends on quenching to freeze in high-temperature microstructure and is sensitive to microscope resolution for detection of phase boundaries. High temperature furnaces are needed along with containers more refractory than the material under study, and interaction of the material with furnace atmosphere is a concern at high temperatures.

In the present work, a novel laser melting technique[12,13,14] for thermal analysis has been applied to the study of ZrC. The essence of the technique is the use of laser pulse heating to melt the material's surface and optical pyrometry to monitor temperature and detect thermal arrests indicative of phase changes. Simultaneously, the reflected light of a second laser from the sample surface is monitored; rapid oscillations in the reflected light signal (RLS) indicate the presence of a liquid on the surface, independently determining solidus and liquidus temperatures.

The laser melting technique offers several benefits over traditional methods for studying refractory materials. Extremely high temperatures are achieved rapidly (ms timescale), with laser pulse profile controllable for optimal heating and cooling rates. Sample vaporisation and oxidation are minimised by using inert gas atmospheres in a pressure vessel. The laser heats a localised region of the sample surface, so reaction with containers is avoided. The sample remains intact, allowing recording of the cooling curve, repeated melting experiments on the same sample, and recovery for post-melting microstructural and chemical analysis.

One concern is the accuracy of optical temperature measurements at extremely high temperatures (>3000 K). Measurements must be made on blackbody specimens, or else material emissivity must be accurately known at the temperatures of interest (which is difficult, especially for the liquid phase). In addition, thermodynamic equilibrium is compromised due to the ultra-fast heating and cooling. Localised melting of the surface means that heat diffuses to the unmelted bulk, and latent heat exchanges are usually not detectable during the heating phase; the RLS and cooling phase of the thermogram provide the bulk of the phase transition information. Finally, variation of power across the laser beam gives rise to variation in the heat treatment temperature across the sample surface; precise alignment of pyrometers and laser beam must be performed.

EXPERIMENTAL

The powder used as raw material was commercial zirconium carbide (ABCR, Germany) with manufacturer-reported 3.9 μm mean particle size and 99.19% purity (1.9 wt% Hf, 0.3553 wt% N, and 0.1991 wt% O). Powder size was determined in this work using a Malvern Mastersizer laser diffraction particle size analyser and by observation in LEO 1525 FEGSEM and JEOL JSM-840 scanning electron microscopes. Powder morphology and compositional homogeneity were also qualitatively studied by SEM.

Phase identification by X-ray diffraction was performed with a Philips PW 1729 X-ray Generator and PW1710 Diffractometer Control, operating at 40 kV and 40 mA, using Ni-filtered Cu K_α radiation. Powders were scanned between 2θ angles of 20-100°, with a scanning step of 0.04° and 2 s per step. Peak positions were calibrated with an external Si standard. The International Centre for Diffraction Data (ICDD) Powder Diffraction File (PDF) was consulted for standard peaks of potential phases. Lattice parameter was calculated based on at least 8 reflections for each specimen and approximation of systematic error using the Nelson-Riley function[15]; lattice parameter extrapolated to 2θ = 180° is reported.

Chemical composition was determined by combustion in a Thermo Electron Flash EA 1112 elemental analyser (for carbon) and inert gas fusion in an Eltra ONH 2000 (for oxygen and nitrogen); zirconium was taken to comprise the balance. Further quantitative analysis for Hf and trace metal content is planned. The results of these analyses were combined to determine stoichiometry.

Spark Plasma Sintering of Zirconium Carbide

Zirconium carbide powders were densified using a spark plasma sintering (SPS) furnace (FCT Systeme, Germany). Powder (6-7 g) was packed in a 20 mm diameter graphite die lined with graphite foil and pre-compacted in a laboratory press. A carbon foam blanket surrounded the die during sintering to reduce heat loss. Temperature was monitored by optical pyrometry into a hollow in the top punch, measuring as close as practical to the sample. All sintering was carried out under vacuum (0.06 mbar).

Four sintering programmes were tested, as depicted in Figure 2. Soak temperatures between 1900 and 2180°C and uniaxial pressures of 40 to 100 MPa were tested. Load was applied while temperature was 1000°C or greater, and ramp rate was 100 or 200°C min^{-1}. The high-temperature soak was terminated manually after piston speed dropped to zero. The samples sintered at 1900 and 2000°C required 6-11 min for densification to terminate, but two samples sintered at 2100 and 2180°C were held for a full 30 min to study the effect on grain coarsening. Cooling from the soak temperature to 1000°C was controlled at -200°C min^{-1} and water cooling brought the furnace to room temperature after approximately 10 min.

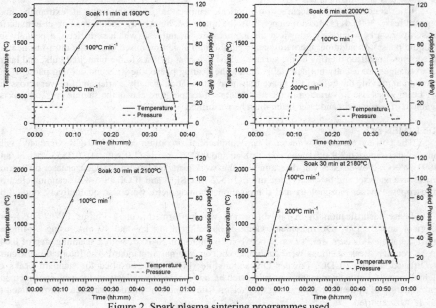

Figure 2. Spark plasma sintering programmes used.

Density of sintered pellets was determined using the Archimedes principle. For ceramography, cross-sections parallel to and transverse to the pressing direction were cold-mounted, diamond-polished to 1 μm surface finish, and etched using a 1:1:1 mixture of HF, HNO$_3$, and distilled water. Microstructure was examined using reflected light microscopy (pre- and post-etching), SEM, and TEM. Grain size was determined by the linear intercept method[16] using ImageJ software[17]. Lattice parameters of sintered samples were measured and compared to those of powders. All sintered samples were presumed to be of comparable composition, and samples were sectioned for laser melting experiments from the SPS sample heated to 2100℃ for 30 min at 40 MPa as well as from the sample heated to 2000℃ for 6 min at 70 MPa.

Laser-Induced Heating and Melting: A Novel Thermal Analysis Technique

A schematic of the laser melting apparatus is depicted in Figure 3. Samples were held in a graphite fixture within a pressure vessel fitted with a sapphire window. All experiments were carried out under Ar gas (2 bar).

A 4.5 kW Nd:YAG CW laser was used to deliver short pulses (70 to 400 ms, peak power 990 to 3800 W) to the sample surface; beam sizes of 3 and 8 mm diameter were used. Several laser profiles were tested to identify best practices for obtaining thermograms with reproducible and well-defined features. Experiments using longer pulses and/or larger diameter beam were used to melt more material and produce a more pronounced thermal arrest upon cooling. A second low power (1 W) Ar⁺ laser (λ = 514 nm) was focused on the same location.

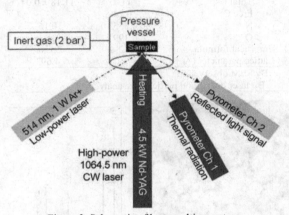

Figure 3. Schematic of laser melting system.

Sample surface temperature was measured with a high-speed two-channel pyrometer. The first channel was operated at λ = 644 nm and calibrated against standard lamps for brightness temperature measurement; the second channel was used to detect the reflected light of the Ar⁺ laser. Brightness temperature was converted to true temperature taking into account the transmittance of the window and the sample emissivity. Values for ZrC emissivity (λ = 0.65 μm) are reported[18] to lie in the range 0.5 to 0.8, decreasing with temperature; pending in-house measurement of emissivity of the present samples, a value of 0.6 was adopted as the best approximation for the high-temperature emissivity of ZrC.

For congruently melting compositions, a freezing plateau in the cooling curve corresponds to the melting/freezing temperature. Commercially-available ZrC is near-stoichiometric and a non-congruently melting composition; a freezing inflection is bounded by distinct liquidus and solidus temperatures. Using the reflected light signal, the onset of oscillations in the RLS – the first formation of liquid – is used to determine the solidus temperature. The termination of oscillations – the formation of a solid crust atop the melt – is used to determine the liquidus temperature.

RESULTS AND DISCUSSION

Properties of the starting powder as reported by the manufacturer and as measured in this work are reported in Table I. A backscattered electron micrograph shows powder size of $0.5 - 10$ μm and equiaxed particles with mixed rounded and angular morphology in Figure 4. X-ray powder diffraction (Figure 5) shows only cubic ZrC peaks with lattice parameter a = 4.693 Å; literature values [3] closer to 4.698 Å are most likely based on purer material.

Table I. Properties of As-Received Zirconium Carbide (ABCR, Germany)

	Manufacturer	Present Work
Mean particle size (μm)	3.9	0.5-10 [*] 6.8 ± 2.1 [**]
Composition (wt%)		
Zr+Hf	88.03	-
Hf	1.9	-
Total C	11.4	11.18 ± 0.04
Free C	0.24	-
N	0.3553	0.16
O	0.1991	0.51
Empirical formula	$ZrC_{0.97}$	$ZrC_{0.96}$
Lattice parameter (Å)	-	4.693

[*] By SEM observation
[**] By laser diffraction particle size analysis

Figure 4. Backscattered electron image of ZrC powder (ABCR, Germany), particle size $0.5 - 10$ μm.

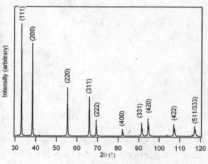

Figure 5. X-ray diffractogram of starting powder. All reflections correspond to cubic ZrC (a = 4.693 Å).

As-Sintered Ceramics

Density, grain size, and lattice parameter of the sintered ceramics are summarised in Table II. Electron micrographs of sintered microstructures are shown in Figures 6 and 7. All samples were densified to greater than 96% relative density (ρ_{th} = 6.62 g cm^{-3}) with the highest density above 99%. The lattice parameters of all samples were slightly lower than that of the starting powder. As lattice parameter correlates both with C:Zr ratio and with oxygen/nitrogen impurity content, shift in lattice parameter may be due to modification of the composition during the sintering process. Sintered grain size of all samples exceeds the powder size range, indicating grain coarsening upon sintering. Samples held at temperature for longer show a greater degree of coarsening compared with those sintered just long enough for piston speed to drop to zero. Higher ramp rate correlated with lower density and high porosity. The lowest temperature, despite combining with the highest load, resulted in the lowest density.

Table II. Characteristics of As-Sintered Ceramics

Sintering Temperature (°C)	Ramp Rate (°C min⁻¹)	Applied Load (MPa)	Soak Time (min)	Archimedes Density (g cm⁻³)	Nominal Grain Diameter (μm)	Lattice Parameter (Å)
1900	100	100	11	6.41	7.5 - 8	4.691
2000	100	70	6	6.59	7 - 9	4.692
2100	100	40	30	6.61	21 - 31	4.692
2180	200	50	30	6.49	17 - 22	4.691

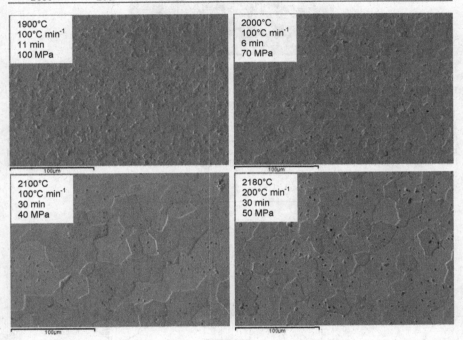

Figure 6. Secondary electron images (same scale) of ZrC microstructure for 4 SPS programmes.

Chemical analysis of the sintered samples is pending, as it is hoped to determine if any composition shift occurs during sintering, as well as to obtain accurate composition for analysis of laser melting results.

Based on these results, a sintering programme of 2000℃, 6 –8 min soak, and 50 MPa was chosen for sintering laboratory-synthesised powders of various C:Zr ratio for a study of thermophysical properties as a function of stoichiometry, currently underway.

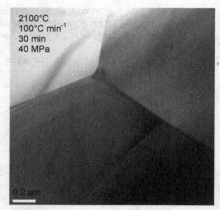

Figure 7. Bright field TEM image of densified ZrC.

Laser Melting Thermal Analysis and Post-Melting Analysis of Samples

A typical experimental thermogram for commercial zirconium carbide is shown in Figure 8. The laser pulse used 3 mm beam diameter, 990 W maximum power, ramped to peak power in 50 ms, and held a further 20 ms. Features identifiable from the thermogram include temperature rise with onset of laser pulse, rapid temperature drop following laser power shut-off, and thermal arrest inflection during freezing. The thermal arrest inflection and oscillations in the differentiated reflected light signal confirm that this sample melted. RLS aided determination of solidus and liquidus temperatures.

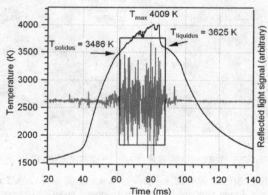

Figure 8. Laser melting thermogram for commercial ZrC specimen
(beam diameter 3 mm, peak power 990 W, ramp time 50 ms, hold 20 ms).

Samples were recovered for post-melting analysis. Secondary electron images of one melted surface are shown in Figure 9. The molten spot is 1 mm in diameter, a small part of the original sample surface. At higher magnification, dendrites are seen in the centre of the melt. XRD of the melted material shows cubic ZrC peaks with some preferred orientation versus powder diffraction peaks.

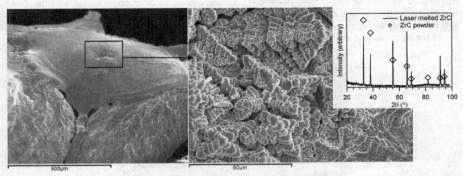

Figure 9. Surface of laser-melted ZrC (peak power 990 W, pulse length 70 ms, beam diameter 3 mm), with dendritic microstructure in melt centre (inset). XRD of melted material shows cubic ZrC peaks.

Short laser pulses with a 3 mm beam size tended to result in sample breakage and loss of RLS signal, hindering analysis of thermograms. Slower pulses (ramped to peak power in 150 ms, held at peak power a further 250 ms) using an 8 mm beam size were employed to melt a larger area of the sample surface as well as to reduce the occurrence of sample breakage. Figure 10 shows thermograms from these experiments; here the laser power was varied to determine power needed to melt the sample.

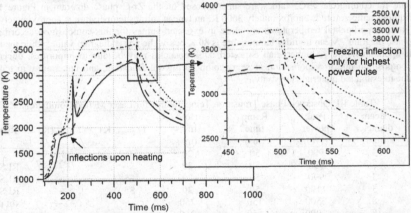

Figure 10. Thermograms for laser melting experiments using larger beam (8 mm diameter) and longer pulses (ramp time 150 ms, hold 250 ms). Inset shows magnification of cooling curve and freezing inflections. Unusual inflections between 1800 – 2000 K upon heating are observed.

A freezing inflection was only observed for the highest power pulse; lower power pulses failed to reach the melting temperature. No eutectic transition was detected in the cooling flank of the curve. In the heating flank, inflections were observed between 1800 - 2000 K. Such inflections have been observed during laser melting of other materials[19] and potentially reflect an emissivity change in this temperature range. The temperature range does not correspond to other phase changes in the Zr-C system (i.e. melting of hcp Zr at 2125 K or ZrC-C eutectic at 3200 K). The source of these inflections will be explored in future experiments. A secondary electron image of the sample surface (Figure 11) shows the entire surface melted.

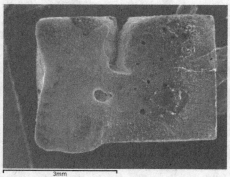

3mm

Figure 11. SEM image of sample melted using larger beam (8 mm diameter) and longer pulses (ramp time 150 ms, hold 250 ms). Entire surface is melted.

Table III summarises phase transition temperatures observed from laser melting experiments to date for commercial ZrC; values are superimposed on the Zr-C phase diagram in Figure 12. Mean liquidus temperature is approximately 3600 K, and mean solidus temperature is approximately 3450 K. These experimental temperatures are close to expected values for this composition according to the phase diagram[6]. Mean liquidus temperature is 80 K below the phase diagram value, while mean solidus temperature is 50 K higher than expected. Discrepancies may be due to impurities, uncertainty in composition due to vaporisation or segregation during solidification, and uncertainty in temperature due to estimated material emissivity.

Table III. Observed Phase Transition Temperatures for Laser Melted Commercial ZrC

Beam diameter (mm)	Peak power (W)	Ramp time (ms)	Hold time (ms)	$T_{liquidus}$ (K)	$T_{solidus}$ (K)	Remarks
3	990	50	20	3625	3486	
3	990	50	20	-	-	RLS Lost
3	990	50	20	3598	-	RLS Lost
3	1170	50	20	3595	-	RLS Lost
8	2500	150	250	-	-	Not melted
8	3000	150	250	-	-	Not melted
8	3500	150	250	-	-	Not melted
8	3800	150	250	3614	3415	
			Mean	3608 ± 14	3451 ± 50	

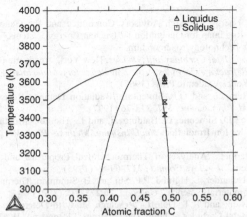

Figure 12. Melting temperatures for commercial ZrC (49 at% C) determined in this work plotted against the high temperature portion of the Zr-C phase diagram.

CONCLUSIONS AND FURTHER WORK

In the spark plasma sintering of ZrC, high temperature was more instrumental in densification than was high pressure. Short process times reduced grain coarsening, but ramping to temperature at too fast a rate resulted in trapped porosity which was difficult to remove by prolonged heating.

In this work, the use of pulsed laser melting has been demonstrated to be a viable technique for probing the extremely high temperature phase transitions observed for zirconium carbide and other refractory ceramics. Temperatures well above the melting temperature of ZrC were obtained rapidly under containerless conditions, and analysis of the resulting thermograms yielded experimental solidus and liquidus temperatures within 50-80 K of predicted values.

Further work will seek to refine and optimise the technique for accurate property measurement. In-house measurement of ZrC emissivity is planned, as is quantitative chemical analysis of sintered and melted samples. It would be beneficial to study the effect of non-congruent vaporisation on composition due to melting experiments. Microstructure of melted samples will be studied in conjunction with thermal analysis results. It will be important to understand the degree to which phase segregation upon solidification affects the analysis. The processing and laser melting of various compositions in the Zr-C system will be reported in a future publication. The results will contribute to the literature on ZrC thermophysical properties and phase transitions.

ACKNOWLEDGMENTS

This work was carried out as part of the Towards a Sustainable Energy Economy (TSEC) programme Keeping the Nuclear Option Open (KNOO) and as such we are grateful to the UK Engineering and Physical Sciences Research Council for funding under grant EP/C549465/1. This work was also supported by the Overseas Research Students Awards Scheme, the Pottery Mechanics Institute Trust Fund, the Institute of Materials, and the Royal Academy of Engineering. Work was carried out under Collaboration Agreement no. 30966 between Imperial College London and the European Commission Joint Research Centre Institute for Transuranium Elements (ITU). The authors wish to thank Dr. Rudy Konings of ITU for fruitful discussions.

REFERENCES

[1]U.S. DOE Nuclear Energy Research Advisory Committee and the Generation IV International Forum, "A Technology Roadmap for Generation IV Nuclear Energy Systems," (2002), Accessed 2008 at http://www.gen-4.org/Technology/roadmap.htm.

[2]L. E. Toth, *Transition Metal Carbides and Nitrides*, New York, Academic Press (1971).

[3]E. K. Storms, *The Refractory Carbides*, Vol. 2 in *Refractory Materials: A Series of Monographs*, J. Margrave, ed., New York, Academic Press (1967).

[4]A.I. Gusev, A.A. Rempel, and G.P. Shveikin, "Radiation Damage Resistance of Materials and Nonstoichiometry," *Dokl. Phys. Chem.*, **357**, 373-76 (1997).

[5]D. Gosset, M. Dollé , D. Simeone, G. Baldinozzi, and L. Thomé , Structural Behaviour of Nearly Stoichiometric ZrC Under Ion Irradiation, *Nucl. Instrum. Methods Phys. Res., Sect. B*, **266**, 2801-05 (2008).

[6]A. Fernandez-Guillermet, "Analysis of Thermochemical Properties and Phase Stability in the Zirconium-Carbon System," *J. Alloys Compd.*, **217**, 69-89 (1995).

[7]J.O. Andersson, T. Helander, L. Hǎlund, P.F. Shi, and B. Sundman, Thermo-Calc and DICTRA, Computational tools for materials science," *Calphad*, **26** 273-312 (2001).

[8]R.V. Sara, C.E. Lowell, and R.T. Doloff, "Research Study to Determine the Phase Equilibrium Relations of Selected Metal Carbides at High Temperatures," Report no. WADD-TR-60-143, Part 4, National Carbon Company, Parma, Ohio (1963).

[9]R. V. Sara, "The System Zirconium-Carbon," *J. Am. Ceram. Soc.*, **48**, 243-47 (1965).

[10]E. Rudy, D. P. Harmon, and C. E. Brukl, "Ternary Phase Equilibria in Transition Metal Boron-Carbon-Silicon Systems, Part 1. Related Binary Systems, Volume 2. Ti-C and Zr-C System," Report no. AFML-TR-65-2, Aerojet-General Corp., Sacramento, CA (1965).

[11]E. Rudy and G. Progulski, "Ternary Phase Equilibria in Transition Metal-Boron-Carbon-Silicon Systems, Part 3. Special Experimental Techniques, Volume 2. A Pirani-Furnace for the Precision Determination of the Melting Temperature of Refractory Metallic Substances," Report no. AFML-TR-65-2, Aerojet-General Corp., Sacramento, CA (1967).

[12]D. Manara, C. Ronchi, M. Sheindlin, M. Lewis, and M. Brykin, "Melting of Stoichiometric and Hyperstoichiometric Uranium Dioxide," *J. Nucl. Mater.*, **342**, 148-63 (2005).

[13]D. Manara, M. Sheindlin, W. Heinz, and C. Ronchi, "New Techniques for High-Temperature Melting Measurements in Volatile Refractory Materials Via Laser Surface Heating," *Rev. Sci. Instrum.*, **79**, 113901 (2008).

[14]C.A. Utton, F. De Brucker, K. Boboridis, R. Jardin, H. Noel. C. Gué neau, and D. Manara, Laser Melting of Uranium Carbides," Proceedings of the E-MRS 2008 Spring Meeting Symposium N on Nuclear Materials, *J. Nucl. Mater.* (2009) In press.

[15]B.D. Cullity, *Elements of X-Ray Diffraction*, 2nd ed., Reading, MA, Addison-Wesley (1978).

[16]ASTM E112, "Standard Test Method for Determining Average Grain Size," ASTM International, West Conshohocken, PA.

[17]M.D. Abramoff, P.J. Magelhaes, and S.J. Ram, "Image Processing with ImageJ," *Biophotonics International*, **11**, 36-42 (2004).

[18]T.E. Zapadaeva, V.A. Petrov, and V.V. Sokolov, "Emissivity of Stoichiometric Zirconium and Titanium Carbides at High Temperatures," *High Temp.*, **19**, 313-20 (1981).

[19]D. Manara, personal communication (2008).

Nuclear Fuels
and Wastes

DEVELOPMENT AND TESTING OF A CEMENT WASTE FORM FOR TRU EFFLUENT FROM THE SAVANNAH RIVER SITE MIXED OXIDE FUEL FABRICATION FACILITY

A.D. Cozzi and E.K. Hansen
Savannah River National Laboratory
Aiken, SC 29803

ABSTRACT

The Savannah River Site Mixed Oxide Fuel Fabrication Facility will generate an effluent stream that consists of small quantities of americium in a nitric acid solution. The Waste Solidification Building is designed to solidify this and other effluent streams from the process.

A cementitious waste form was developed that will solidify neutralized TRU effluent and produce a waste form that will meet the criteria for the Waste Isolation Pilot Plant.

INTRODUCTION

The Savannah River Site Nuclear Nonproliferation Programs is in the final design stage of the Waste Solidification Building (WSB). This facility will provide treatment and solidification of the three radioactive liquid waste streams generated by the Pit Disassembly and Conversion Facility (PDCF) and the Mixed Oxide (MOX) Fuel Fabrication Facility (MFFF). The WSB will independently treat a transuranic (TRU) high alpha (HA) waste stream and two low-activity level waste (LLW) streams. Cementation has been selected as the solidification method for both the high-alpha (TRU-HA) and the LLW waste streams generated in the PDCF and MFFF. This work will address only the TRU-HA waste stream because it is the waste stream with the most processing constraints.

The HA waste will be a moderately acidic aqueous stream generated in the following processes: the MOX acid recovery and recycle processes; the alkaline treatment process; and the aqueous purification process. This waste will also includes rinse water from equipment and drain flushes and is expected to have a pH <1. The various waste streams will be evaporated to reduce the volume of the waste. Prior to the cementitious process, the waste streams will be neutralized using concentrated NaOH solution. Additional NaOH solution will be added to make the streams slightly caustic, targeting a pH of 12.

In the WSB process, neutralized waste solutions will be added to stainless steel 55 gallon drums containing cementitious dry materials. The waste solutions are blended with the dry solids using mechanical agitation and upon completion of mixing the drums are secured and vented. The drums will then be stored until disposition at the appropriate waste facility.

The objective of this work was to:
- Produce a simulant for the waste
  - Components, concentrations and substitutions
- Develop prospective waste forms to meet processing requirements
  - In drum mixing
  - Contact handle drums (thermal and radiation)
  - Throughput (8 drums/24 h)
- Develop waste form to meet storage requirements for final disposal at WIPP
  - Free water
  - Hydrogen generation/dissipation

EXPERIMENTAL

Simulant Development

Based on the anticipated feed stream from the MFFF, a salt solution was prepared to simulate the WSB evaporator bottoms. Table I is the simulant of the acidic product. For these tests, erbium was used as a non-radioactive surrogate for americium.[1] The acidic solution is neutralized using 50 wt% NaOH and then further additions are made to adjust the pH to 12. The resulting salt solution contains approximately 37 wt% solids, almost all of the solids are dissolved in the solution with less than two percent undissolved solids.

Table I. Batch Sheet to Make 1-Liter of the Acidic HAW Simulant

Chemical Name	Compound	Target (g)
Nitric Acid	HNO_3 (69.9%)	666.178
Erbium Nitrate Pentadydrate	$Er(NO_3)_3 \cdot 5H_2O$	0.925
Gallium Nitrate Hexahydrate	$Ga(NO_3)_3 \cdot 6H2O$	5.479
Sodium Nitrate	$NaNO_3$	14.024
Silver Nitrate (100%)	$AgNO_3$	5.480
Barium Nitrate	$Ba(NO_3)_2$	0.609
Calcium Nitrate Tetrahydrate	$Ca(NO_3)_2 \cdot 4H_2O$	22.271
Cadmium Nitrate Tetrahydrate	$Cd(NO_3)_2 \cdot 4H_2O$	0.165
Potassium Nitrate	KNO_3	15.385
Magnesium Nitrate Hexahydrate	$Mg(NO_3)_2 \cdot 6H_2O$	23.308
DI water	H_2O	522.825

Testing encompassed a range of water-to-cement ratios (w/c) as well a variety of density modifiers and loadings of the density modifiers. The object of the testing was to prepare a waste from that would provide sufficient self shielding to be contact handled in the facility. Additional testing was performed to identify surrogates for the costly components gallium and silver, and the hazardous components barium and cadmium (silver is also a hazardous component). This is necessary to support full-scale testing. The resulting acidic simulant is shown in Table II.

Table II. Batch Sheet to Make 1-Liter of the Acidic HAW Simulant for Full-Scale Testing

Chemical Name	Compound	Target (g)
Nitric Acid (69.9%)	HNO_3 (69.9 wt%)	662.2
Aluminum Nitrate Nonahydrate	$Al(NO_3)_3 \cdot 9H_2O$	5.6
Sodium Nitrate	$NaNO_3$	14.0
Copper Nitrate Hemipentahydrate	$Cu(NO_3)_2 \cdot .5H_2O$	7.5
Calcium Nitrate Tetrahydrate	$Ca(NO_3)_2 \cdot 4H_2O$	23.0
Potassium Nitrate	KNO_3	15.4
Magnesium Nitrate Hexahydrate	$Mg(NO_3)_2 \cdot 6H_2O$	23.3
DI Water	H_2O	514.8

Waste Form Preparation

Individual simulant batches were made based on the analysis of the simulant in Table II. Four tests, two simulants at ambient temperature to prepare drums filled to either 65 or 75% full and two at an elevated temperature, 43 °C, to represent worst case chiller performance in the WSB.

The grout batches all consisted of a water to cement ratio of 0.30, and a density modifier (zircon flour) to cement ratio of 1/12. The quantity of simulant made was based on using the densities and mass fractions of the various components (simulant, cement, and zircon flour) and volume additively to determine the final grout density.

The salt solution batches were made with commercial grade chemicals. The acid solution was initially made, physically characterized and then neutralized and adjusted to a pH of 12 ± 0.3 using a concentrated (~51 wt %) NaOH solution. The resulting salt solutions were analyzed for density and solids. Based on the total solids in each batch, the quantity of Portland Type I/II (GU-general use) cement and zircon flour were then determined.

Grout Mixer Operation

An IMER grout mixer was selected based on its ability to process at least 55 gallons of grout and that a standard 55 gallon drum could be slid under the discharge of the mixer. The mixer was modified with leg extensions that provided a more stable configuration for drum fill. The side view of the elevated mixer and the internals of the mixer are shown in Figure 1. The agitator rotates at 22 revolutions per minute and is belt driven using a 5 horsepower electric motor.

Figure 1. IMER Mortarman360 Grout Mixer.

The HA waste simulants were homogenized prior to pumping it into the mixer. For the elevated temperature runs, the simulants were preheated to at least 43 °C prior to pumping it into the mixer. The temperatures of the elevated simulants were measured after they were pumped into the mixer to obtain a starting simulant temperature. The dry materials were maintained at room temperature.

Instrumentation and Test Setup

To measure the temperature of the curing grout, a thermocouple tree was installed in the 55 gallon drum, measuring the center, 1/3 radial, 2/3 radial, and near the wall of the drum positions. In each of these radial positions, a T-type thermocouple was located at five axial positions, evenly spaced between the top and bottom. A total of 20 grout temperature measurements were obtained, per test. Figure 2 is a schematic of the thermocouple placements on the tree, for a 75% full drum condition. Figure 3 shows a thermocouple tree, free standing and installed in a 55 gallon drum. The frame of the tree was made of stainless steel. The thermocouple trees were sized to fit the pour size, e.g. 65% or 75% full drums. The external skin temperatures of the 55 gallon stainless steel drums were measured at four axial positions evenly spaced out between the top and bottom of the active grout pour. The skin thermocouples design is integrated with a cement adhesive that permitted the thermocouple to be attached directly to the stainless steel drum, with little to no impact on heat flow. The temperature, humidity, and dew point in the vapor space in the 55 gallon drums were measured using a DLP-design-TH1 sensor. A single standard 55 gallon drum lid was modified, where hermitically sealed electrical connections and a vent line were installed. This modified lid was used in each test.

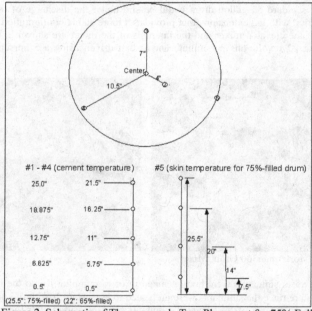

Figure 2. Schematic of Thermocouple Tree Placement for 75% Full Drum.

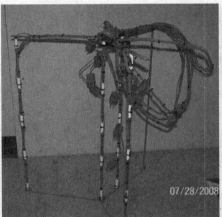
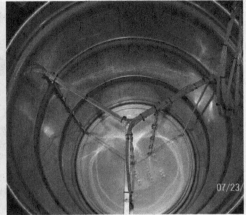

Figure 3. Thermocouple Tree Freestanding and Placed for 75% Full Drum.

RESULTS

Mixer Operations

First, the salt solution was first added to the mixer and the agitator energized. After all the cement and zircon flour was added, the grout was allowed to mix for ten minutes prior to discharging. The resulting grout was visually homogenous, very flowable in both the mixer and as it discharged the mixer into the 55 gallon drum.

Figure 4 shows the initial fill and instrument connecting activities for the 65% full ambient temperature run. Figure 5 shows this drum after two days of curing. Salt crystals are present on the surface with no bleed water. In all of the tests, there was no observable bleed water after 24 hours of curing. During all these runs, there were no observable pressure increases in the drum vapor space.

Thermal Properties

The heat flow (or heat generation rate) was measured using isothermal calorimetry from a sample of the simulated HA waste grout collected during the pouring of the 65% full, elevated temperature salt solution run. Normal sample preparation and instrument setup leads to a delay of 30 to 45 minutes from grout collection to the start of data collection. The data were collected by maintaining the sample at 25 °C, in the TAM isothermal calorimeter. Figure 6 illustrates the heat flow of the sample after pouring (normalized to the mass of grout) due to the heat of hydration of the curing cement. Experience indicate that as the temperature of the grout is increased, either by using elevated temperature salt solution or due to temperature rise from the hydration reaction, the thermodynamics (area under the heat flow curve) do not change, but the kinetics (shape of the curve) change. As the temperature of the grout rises, the peak occurs sooner and results in a larger maximum heat flow. The measured specific heat on the same sample was determined to be $1.6 \, {}^{kJ}\!/_{kg \cdot K}$.

Figure 4. HA Waste Grout 65% Full Ambient Temperature Run.

Figure 5. HA Waste Grout 65% Full Ambient Temperature Run – After Curing
For 2 Days.

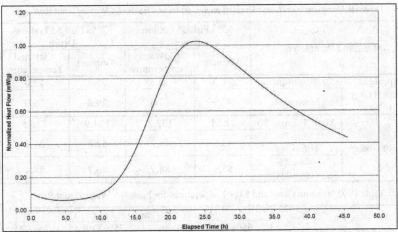

Figure 6. Normalized Heat Flow for HA Waste Grout Measured at 25 °C.

Temperature Profiles

Table III contains the minimum and maximum ambient and vapor space temperature and vapor space humidity. The relative humidity in the vapor space of the 55 gallon drums never exceeded 90% and the maximum temperature never exceed 140 °F. The maximum grout and skin temperatures for each run are provided in Table IV. The maximum temperature in the grout exceeded the boiling point of water for all the runs. There were no observable water vapor paths from the internal regions of the grout to the grout surface. The maximum skin temperature exceeded 140 °F for all cases, except for the 65% full elevated temperature run. Table V contains the times at which the maximum temperature occurred for each temperature measurement location. Appendix A contains the measured temperature distributions and vapor space humidity for all the runs. The 75% full room temperature run has a data gap in the recorded data and is due to a power interruption.

Table III. Ambient Room and Vapor Space Conditions Each HA Waste Grout Run

Measured Parameter		65% Full 55 Gallon Drum		75% Full 55 Gallon Drum	
		Ambient	Elevated Temperature	Ambient	Elevated Temperature
Ambient Room Temp (°F)	Minimum	71.0	70.5	72.2	72.0
	Maximum	77.9	87.6	79.8	76.6
Vapor Space	Max. Temp (°F)	122.1	122.2	134.9	136.5
	Min. % Humidity	43.8	46.2	52.8	51.2
	Max. % Humidity	85.9	88.7	86.7	88.4

Table IV. Maximum Grout and Skin Temperatures for Each HA Waste Grout Run

Location in/on 55 gallon drum		Maximum Temperatures (°F)			
		65% Full 55 Gallon Drum		75% Full 55 Gallon Drum	
		Ambient	Elevated Temperature	Ambient	Elevated Temperature
Center	Top	147.2	141.5	179.4	190.3
	2nd from top	210.5	203.4	232.7	234.8
	Midpoint	227.6	222.1	245.9	244.7
	2nd from bottom	220.2	217.1	236.9	233.7
	Bottom	175.6	173.9	186.8	191.2
1/3rd radius	Top	150.5	135.4	172.6	188.4
	2nd from top	207.8	201.1	228.2	230.5
	Midpoint	225.3	219.4	240.2	240.4
	2nd from bottom	218.1	214.1	231.2	229.5
	Bottom	175.8	175.2	183.6	190.1
2/3rd radius	Top	135.2	125.3	169.5	170.1
	2nd from top	182.3	176.6	213.1	209.5
	Midpoint	199.2	191.8	220.9	216.6
	2nd from bottom	192.0	186.1	216.6	210.8
	Bottom	153.3	150.5	174.5	168.4
Inside wall	Top	119.8	121.7	139.0	147.6
	2nd from top	143.8	153.5	167.2	175.4
	Midpoint	155.2	165.9	170.5	172.2
	2nd from bottom	148.4	157.5	161.9	164.2
	Bottom	127.2	125.3	139.8	136.2
Outside skin	Top	140.4	126.5	152.0	150.3
	2nd from top	154.7	139.2	162.1	156.4
	2nd from bottom	154.7	134.5	161.9	163.2
	Bottom	147.2	117.5	145.9	141.7

Table V. Time at Which Maximum Temperature Was Reached in the Grout for Each HA Waste Grout Run

Location in/on 55 gallon drum	Time max temperature occurred after pour (hours)			
	65% Full 55 Gallon Drum		75% Full 55 Gallon Drum	
	Ambient	Elevated Temperature	Ambient	Elevated Temperature
Center	28	23.5	28.5	20.9
1/3rd radius	27	24.0	28	20.5
2/3rd radius	27	24.0	26	19.5
Inside wall	25	23.6	d/l	19.3
Outside skin	25	23.5	d/l	20.8

d/l = data lost

CONCLUSIONS

In all tests, the dry materials were maintained at ambient temperature, prior to mixing. During the 55 gallon drum tests, there was no buildup of internal pressure in the vapor space during any of the tests.

The drum temperatures, other than for the 65% full, elevated temperature run, all had maximum skin temperatures at or above 140 °F. For the 65% full, ambient temperature run, the + 140 °F skin temperature occurred between 22 to 32 hours after the grout was poured. For the 75% full, elevated temperature run, the + 140 °F occurred between 16 to 25 hours. Comparison of the room and elevated runs indicate that the increase in temperature of the salt solution causes the maximum temperature at each measurement point to occur sooner, directly impacting the rate of hydration. The skin temperatures for the elevated temperature runs initially started at a higher temperature than ambient conditions, but cooled prior to reheating. The central locations of the curing grout for the elevated temperature runs did not decrease until the maximum temperature was obtained. This was not true for the grout near the drum surfaces, where temperature first decreased and then increased.

The maximum grout temperature was located at the center of the drum and exceeded 222 °F (105 °C) for all cases. This value exceeds the boiling point of water. Review of the data in Appendix A support the above claims on the impact of elevated salt solution to the internal temperatures of the grout as well as that of the skin temperature. There were additional points of interest, such as the oscillatory data towards the end of the runs for skin and outer most T/Cs, indicating localized hydration reactions occurring.

The composition of the dry materials could also impact the temperature of the final grout. The 65% full, elevated temperature run temperatures were all lower than that of 65% full, ambient temperature run. The primary difference, other than solution temperature, was the sources of Portland cement used in the two runs. Comparison of the 75% full runs showed that even though more Portland cement (of the same source) was used in the ambient temperature run as compared to the elevated temperature run, the elevated temperature run yielded overall higher temperatures. This indicates that starting temperature of the solution is the primary driver, in this case, for temperature distributions.

Inspection of the drum internals, after curing for at least two days, showed condensation had formed on the surfaces of the drum in the vapor space and collected on the surface of the waste form. No bleed was water detected, because the water-to-cement ratio of the grout is such that the waste form is self desiccating and will incorporate additional water. The presence of water on the waste form surface can be determined from the salts present on the surface, such as that shown in Figure 5. The vapor space relative humiditydecreased after the maximum grout temperatures had occurred, indicating that water vapor was being absorbed back into the grout, due to the grout desiccating property.

REFERENCES
[1] R. Villarreal and D. Spall, "Selection of Actinide Chemical Analogues for WIPP Tests," LA-13500-MS, Los Alamos National Laboratory, Los Alamos, NM August 1998.

FRIT OPTIMIZATION FOR SLUDGE BATCH PROCESSING AT THE DEFENSE WASTE PROCESSING FACILITY

Kevin M. Fox, David K. Peeler, Thomas B. Edwards
Savannah River National Laboratory
Aiken, SC 29808 USA

ABSTRACT

The Savannah River National Laboratory (SRNL) Frit Development Team recommended that the Defense Waste Processing Facility (DWPF) utilize Frit 418 for initial processing of high level waste (HLW) Sludge Batch 5 (SB5). The extended SB5 preparation time and desire to avoid a DWPF outage have necessitated the use of a frit that is already included on the DWPF procurement specification. Frit 418 has been used previously in vitrification of Sludge Batch 3 and was the transitional frit to Sludge Batch 4. Paper study assessments predict that Frit 418 will form an acceptable glass when combined with SB5 over a range of waste loadings (WLs), typically 30-41 wt % based on nominal projected SB5 compositions. Frit 418 has a relatively high degree of robustness with regard to potential variation in the projected SB5 composition, particularly when the Na_2O concentration is varied. Frit 418 has not been designed to provide an optimal melt rate with SB5, but is recommended for initial processing of SB5 until experimental testing to optimize a frit composition for melt rate can be completed. Melt rate performance can not be predicted through the paper study assessments and must be determined experimentally.

INTRODUCTION

The objective of this task was to identify a frit for vitrification of high level waste (HLW) Sludge Batch 5 (SB5) at the Savannah River Site (SRS) Defense Waste Processing Facility (DWPF) that:
- is currently listed on the DWPF procurement specification to reduce the time necessary for procurement (due to the short time period available for frit development work),
- will produce acceptable (in terms of repository requirements and DWPF processing constraints) glasses with SB5,
- and is relatively insensitive (based on projected operating windows) to the uncertainty in the SB5 composition projections.

These objectives were met by:
- identifying the best available composition projections for SB5,
- adding any necessary corrections to the composition projections to account for planned caustic additions and blending operations,
- evaluating the operating windows (the range of waste loadings where acceptable glasses are predicted) projected for Frit 418 with these nominal SB5 projections,
- and evaluating the operating windows projected for Frit 418 with variation applied to a bounding SB5 compositional region.

Each of these steps will be described in further detail in the following sections.

SB5 COMPOSITION PROJECTIONS

A series of SB5 composition projections was received from the Liquid Waste Organization (LWO). SRNL also developed projections for SB5 using a model-based approach.[1] The LWO and SRNL projections were based on the heel of material remaining in the feed tank (Tank 40) at the end of Sludge Batch 4 (SB4) processing, the SB5 batch (Tank 51) prior to the transfer to the feed tank, and the SB5 blend in the feed tank after blending with the SB4 heel. None of the projections accounted for the planned addition of caustic to the feed tank to adjust for a decant of excess water. The SB5 blend projections did not account for the addition of the Actinide Removal Process (ARP) stream to SB5 at DWPF. In addition, the final blend ratio of the SB5 batch with the heel in the feed tank to constitute the SB5 blend is uncertain due to the varying sludge usage over the next four months of SB4 processing.

To account for these additions and the blending uncertainties, the LWO and SRNL projected compositions of the Tank 40 heel and the Tank 51 SB5 batch were used as the basis to develop various sludge composition options. First, projections of the Tank 40 heel composition based on the planned addition of caustic were developed by adding 3 wt % Na_2O to the original Tank 40 heel projections and renormalizing the composition to 100 wt %. The composition projections provided by LWO, along with the 3 wt % Na_2O addition to the Tank 40 heel are listed in Table 1.

Table 1. Composition Projections (wt % Calcined Oxides) of the Tank 40 Heel and Tank 51 Provided by LWO, as well as a 3 wt % Na_2O addition to the Tank 40 Heel Projection.

Oxide	Tank 40 Heel	Tank 40 Heel plus 3 wt% Na_2O	Tank 51 (SB5 Batch)
Al_2O_3	26.960	26.013	19.673
BaO	0.077	0.075	0.000
CaO	2.969	2.865	2.258
Ce_2O_3	0.070	0.067	0.000
Cr_2O_3	0.174	0.168	0.000
CuO	0.062	0.060	0.000
Fe_2O_3	30.821	29.737	27.594
K_2O	0.000	0.000	0.036
La_2O_3	0.058	0.056	0.000
MgO	2.927	2.824	1.190
MnO	6.184	5.967	5.828
Na_2O	14.649	17.649	29.494
NiO	1.716	1.656	3.512
PbO	0.064	0.062	0.000
SO_4^{2-}	0.831	0.802	0.373
SiO_2	2.906	2.804	2.217
ThO_2	0.000	0.000	0.000
TiO_2	0.050	0.048	0.000
U_3O_8	9.401	9.071	7.825
ZnO	0.000	0.000	0.000
ZrO_2	0.080	0.078	0.000

The composition projections developed by SRNL, along with the 3 wt % Na_2O addition to the Tank 40 heel are listed in Table 2.

Table 2. Composition Projections (wt% Calcined Oxides) of the Tank 40 Heel and Tank 51 Developed by SRNL, as well as a 3 wt% Na₂O addition to the Tank 40 Heel Projection.

Oxide	Tank 40 Heel	Tank 40 Heel plus 3 wt% Na$_2$O	Tank 51 (SB5 Batch)
Ag$_2$O	0.009	0.009	0.004
Al$_2$O$_3$	25.976	25.040	20.874
BaO	0.078	0.075	0.164
CaO	2.868	2.764	2.366
CdO	0.322	0.310	0.027
Ce$_2$O$_3$	0.068	0.066	0.000
Cr$_2$O$_3$	0.150	0.145	0.088
CuO	0.058	0.055	0.015
Fe$_2$O$_3$	33.075	31.883	28.881
Gd$_2$O$_3$	0.015	0.014	0.026
K$_2$O	0.226	0.218	0.101
La$_2$O$_3$	0.051	0.049	0.000
Li$_2$O	0.041	0.040	0.005
MgO	2.820	2.718	1.253
MnO	5.969	5.754	6.106
Na$_2$O	12.739	15.739	25.090
NiO	1.654	1.594	3.680
P$_2$O$_5$	0.989	0.954	0.125
PbO	0.039	0.038	0.004
PdO	0.001	0.001	0.001
PuO$_2$	0.000	0.000	0.010
Rh$_2$O$_3$	0.015	0.015	0.034
RuO$_2$	0.060	0.058	0.171
SO$_4{}^{2-}$	0.899	0.866	0.539
SiO$_2$	2.795	2.694	2.071
SrO	0.040	0.039	0.080
ThO$_2$	0.000	0.000	0.000
TiO$_2$	0.040	0.039	0.000
U$_3$O$_8$	9.068	8.742	8.191
ZnO	0.005	0.005	0.012
ZrO$_2$	0.079	0.076	0.171

Second, two possible blending ratios were considered for constitution of the SB5 blend.[a] Mass ratios of 25:75 and 30:70 (Tank 40 to Tank 51) were used in blending the Tank 40 heel and Tank 51 SB5 batch compositions, both with and without the caustic addition, using both the LWO and SRNL projections. These factors resulted in eight potential compositions for SB5, as listed in Table 3.

---

[a] The final blend ratio is dependent mainly on the rate of SB4 processing, in DWPF.

Table 3. Potential Compositions of the SB5 Blend Using the LWO and SRNL Projections at Two Blend Ratios, With and Without Caustic Addition to Tank 40.

Tank 40 Source	LWO	LWO	LWO +3 wt % Na$_2$O	LWO +3 wt % Na$_2$O	SRNL	SRNL	SRNL +3 wt % Na$_2$O	SRNL +3 wt % Na$_2$O
Tank 40 Mass Ratio	25	30	25	30	25	30	25	30
Tank 51 Source	LWO	LWO	LWO	LWO	SRNL	SRNL	SRNL	SRNL
Tank 51 Mass Ratio	75	70	75	70	75	70	75	70
Sludge ID	**BS-01**	**BS-02**	**BS-03**	**BS-04**	**BS-05**	**BS-06**	**BS-07**	**BS-08**
Ag$_2$O	0.000	0.000	0.000	0.000	0.005	0.005	0.005	0.005
Al$_2$O$_3$	21.495	21.859	21.258	21.575	22.149	22.405	21.915	22.124
BaO	0.019	0.023	0.019	0.022	0.142	0.138	0.142	0.137
CaO	2.436	2.472	2.410	2.440	2.491	2.516	2.465	2.485
CdO	0.000	0.000	0.000	0.000	0.101	0.116	0.098	0.112
Ce$_2$O$_3$	0.017	0.021	0.017	0.020	0.017	0.020	0.016	0.020
Cr$_2$O$_3$	0.043	0.052	0.042	0.050	0.103	0.106	0.102	0.105
CuO	0.016	0.019	0.015	0.018	0.026	0.028	0.025	0.027
Fe$_2$O$_3$	28.400	28.562	28.130	28.237	29.930	30.139	29.632	29.782
Gd$_2$O$_3$	0.000	0.000	0.000	0.000	0.023	0.023	0.023	0.023
K$_2$O	0.027	0.025	0.027	0.025	0.132	0.138	0.130	0.136
La$_2$O$_3$	0.015	0.017	0.014	0.017	0.013	0.015	0.012	0.015
Li$_2$O	0.000	0.000	0.000	0.000	0.014	0.016	0.013	0.015
MgO	1.624	1.711	1.598	1.680	1.645	1.723	1.620	1.693
MnO	5.917	5.935	5.863	5.870	6.072	6.065	6.018	6.000
Na$_2$O	25.783	25.040	26.533	25.940	22.003	21.385	22.753	22.285
NiO	3.063	2.973	3.048	2.955	3.174	3.072	3.159	3.054
P$_2$O$_5$	0.000	0.000	0.000	0.000	0.341	0.384	0.332	0.374
PbO	0.016	0.019	0.015	0.019	0.013	0.015	0.013	0.014
PdO	0.000	0.000	0.000	0.000	0.001	0.001	0.001	0.001
Pr$_2$O$_3$	0.000	0.000	0.000	0.000	0.000	0.000	0.000	0.000
PuO$_2$	0.000	0.000	0.000	0.000	0.008	0.007	0.008	0.007
Rh$_2$O$_3$	0.000	0.000	0.000	0.000	0.030	0.029	0.029	0.028
RuO$_2$	0.000	0.000	0.000	0.000	0.143	0.137	0.142	0.137
SO$_4$$^{2-}$	0.487	0.510	0.480	0.502	0.629	0.647	0.621	0.637
SiO$_2$	2.389	2.424	2.364	2.393	2.252	2.288	2.227	2.258
SrO	0.000	0.000	0.000	0.000	0.070	0.068	0.069	0.067
ThO$_2$	0.000	0.000	0.000	0.000	0.000	0.000	0.000	0.000
TiO$_2$	0.012	0.015	0.012	0.014	0.010	0.012	0.010	0.012
U$_3$O$_8$	8.219	8.298	8.137	8.199	8.410	8.454	8.329	8.356
ZnO	0.000	0.000	0.000	0.000	0.010	0.010	0.010	0.010
ZrO$_2$	0.020	0.024	0.019	0.023	0.148	0.144	0.148	0.143

Third, the addition of the ARP stream was included for each of the above compositions, resulting in eight more potential compositions for SB5 as listed in Table 4.

Table 4. Potential Compositions of the SB5 Blend Using the LWO and SRNL Projections at Two Blend Ratios, With and Without Caustic Addition, With the ARP Stream Added.

Tank 40 Source	LWO	LWO	LWO +3 wt % Na_2O	LWO +3 wt % Na_2O	SRNL	SRNL	SRNL +3 wt % Na_2O	SRNL +3 wt % Na_2O
Tank 40 Mass Ratio	25	30	25	30	25	30	25	30
Tank 51 Source	LWO	LWO	LWO	LWO	SRNL	SRNL	SRNL	SRNL
Tank 51 Mass Ratio	75	70	75	70	75	70	75	70
Sludge ID	BS-09	BS-10	BS-11	BS-12	BS-13	BS-14	BS-15	BS-16
Ag_2O	0.003	0.003	0.003	0.003	0.007	0.008	0.007	0.008
Al_2O_3	20.741	21.087	20.516	20.817	21.342	21.584	21.127	21.326
BaO	0.023	0.026	0.022	0.026	0.139	0.135	0.139	0.134
CaO	2.365	2.399	2.340	2.369	2.415	2.439	2.391	2.410
CdO	0.000	0.000	0.000	0.000	0.096	0.110	0.093	0.106
Ce_2O_3	0.025	0.028	0.024	0.027	0.024	0.028	0.024	0.027
Cr_2O_3	0.047	0.055	0.045	0.053	0.103	0.106	0.102	0.105
CuO	0.017	0.020	0.016	0.019	0.027	0.029	0.026	0.028
Fe_2O_3	27.600	27.753	27.342	27.444	29.025	29.223	28.752	28.896
Gd_2O_3	0.000	0.000	0.000	0.000	0.022	0.022	0.022	0.022
K_2O	0.031	0.030	0.031	0.030	0.131	0.137	0.129	0.135
La_2O_3	0.017	0.020	0.017	0.019	0.015	0.018	0.015	0.017
Li_2O	0.000	0.000	0.000	0.000	0.013	0.015	0.013	0.014
MgO	1.549	1.632	1.524	1.602	1.567	1.642	1.544	1.613
MnO	5.850	5.867	5.799	5.805	5.992	5.985	5.943	5.926
Na_2O	26.491	25.786	27.204	26.641	22.875	22.288	23.596	23.152
NiO	2.979	2.894	2.965	2.877	3.081	2.985	3.068	2.969
P_2O_5	0.002	0.002	0.002	0.002	0.326	0.367	0.318	0.357
PbO	0.021	0.024	0.021	0.024	0.018	0.020	0.018	0.020
PdO	0.000	0.000	0.000	0.000	0.001	0.001	0.001	0.001
Pr_2O_3	0.003	0.003	0.003	0.003	0.003	0.003	0.003	0.003
PuO_2	0.000	0.000	0.000	0.000	0.007	0.007	0.007	0.007
Rh_2O_3	0.000	0.000	0.000	0.000	0.028	0.027	0.028	0.027
RuO_2	0.006	0.006	0.006	0.006	0.142	0.137	0.142	0.136
SO_4^{2-}	0.606	0.628	0.599	0.619	0.740	0.757	0.732	0.748
SiO_2	2.305	2.338	2.281	2.309	2.172	2.207	2.149	2.179
SrO	0.002	0.002	0.002	0.002	0.068	0.066	0.068	0.066
ThO_2	0.000	0.000	0.000	0.000	0.000	0.000	0.000	0.000
TiO_2	1.311	1.313	1.310	1.312	1.307	1.309	1.307	1.309
U_3O_8	7.973	8.048	7.895	7.954	8.147	8.188	8.072	8.099
ZnO	0.004	0.004	0.004	0.004	0.014	0.013	0.014	0.013
ZrO_2	0.029	0.033	0.028	0.032	0.151	0.146	0.150	0.146

ASSESSMENTS OF FRIT 418 WITH THE NOMINAL SB5 COMPOSITIONS

The 16 potential SB5 compositions described in the previous section (referred to as nominal SB5 compositions) were combined with Frit 418 over a WL interval of 25-60 wt % and evaluated against the DWPF Product Composition Control System (PCCS) Measurement Acceptability Region (MAR) criteria to identify WLs where acceptable glasses are predicted.[2] The results of the Nominal Stage MAR assessment are given in Table 5. In general, the window of available WLs with Frit 418 is quite good, with all of the nominal compositions predicted to form acceptable glasses from 25 to 41% WL (with several of the individual systems having even wider WL windows). Upper WLs are limited by predictions of low viscosity (lowv), high liquidus temperature (T_L), or nepheline crystallization (Neph).

Table 5. MAR Assessment Results for the Nominal SB5 Compositions with Frit 418.

Sludge ID	Acceptable WLs (%)	Upper Limiting Constraint(s)
BS-01	25-43	lowv
BS-02	25-44	lowv, Neph
BS-03	25-42	lowv
BS-04	25-43	lowv
BS-05	25-44	T_L
BS-06	25-43	T_L
BS-07	25-45	T_L, Neph
BS-08	25-44	T_L
BS-09	25-42	lowv
BS-10	25-43	lowv
BS-11	25-41	lowv
BS-12	25-42	lowv
BS-13	25-46	T_L, lowv, Neph
BS-14	25-45	T_L
BS-15	25-45	lowv, Neph
BS-16	25-45	Neph

The results of the MAR assessment with the nominal SB5 compositions show that Frit 418 is robust to these variations for the SB5 composition, including differences between the LWO and SRNL projections, the addition of caustic, varying blends of Tank 40 and Tank 51, and the addition of the ARP stream. Frit 418 appears to be particularly robust to a range of Na_2O concentrations for SB5. The WL windows over which the glasses are predicted to be acceptable are generally limited by process-related constraints (lowv and T_L). Five of the WL windows are limited by predictions of nepheline crystallization (which can impact durability of the glass). However, nepheline is only predicted to form at WLs that are significantly higher (>44% WL) than those likely to be targeted by DWPF.

VARIATION STAGE ASSESSMENT OF FRIT 418 BOUNDING POTENTIAL SB5 COMPOSITIONS

A Variation Stage assessment was next performed to further demonstrate the ability of Frit 418 to accommodate variation in the composition of SB5. The following strategy was developed to apply variation to the compositional region bounding the series of potential SB5 compositions. First, the minimum and maximum concentrations of each component across all 16 of the potential SB5 compositions were determined. Then, for each of the major components (Al_2O_3, Fe_2O_3, Na_2O and U_3O_8), the minimum concentration was reduced by 7.5% and the maximum concentration was increased by 7.5%. For each of the minor components (CaO, MgO, MnO, NiO, SiO_2 and TiO_2), 0.5 wt % was deducted from the minimum concentration and 0.5 wt % was added to the maximum concentration. The remaining components were grouped into a category called 'Others'. The sum of the mean concentrations of each of the remaining components over the 16 potential SB5 compositions was taken as the concentration of 'Others'. A variation of +/- 0.5 wt% was then applied to the concentration of 'Others'. The resulting compositional space defined for the Variation Stage assessment is given in Table 6.

Table 6. Concentration Ranges for Individual Components in the SB5 Composition as Defined for the Variation Stage Assessment.

Oxide	Minimum (wt %)	Maximum (wt %)
Al_2O_3	18.977	24.085
CaO	1.840	3.016
Fe_2O_3	25.292	32.400
MgO	1.024	2.223
MnO	5.299	6.572
Na_2O	19.781	29.245
NiO	2.377	3.674
SO_4^{2-}	0.380	0.857
SiO_2	1.649	2.924
TiO_2	0.000	1.813
U_3O_8	7.302	9.088
Others	0.284	1.284

Table 6 provides the framework around which the Variation Stage assessment was conducted. A sludge composition is in the region defined in Table 6 if its concentration for each oxide is within the minimum and maximum interval for that oxide (e.g., the Al_2O_3 concentration in the sludge is between 18.98 and 24.09 wt %) and the sum of the concentrations of all of the oxides in the sludge equals 100 wt %. Such a composition is a mixture of oxides at concentrations that correspond to one of the possible compositions for SB5 as defined by Table 6. Algorithms are available in statistical software packages[a] to generate the compositions that are the bounding "corner points" of the compositional region defined by Table 6. The bounding compositions generated by the software are called the extreme vertices (EVs) of the compositional region.

A select set (WL interval of 28-39%) of the Variation Stage MAR assessment results is shown in Table 7. For each WL, the number of acceptable and non-acceptable EVs is shown, along with the constraint(s) limiting the non-acceptable EVs.

Table 7. Portion of the Variation Stage MAR Assessment Results for SB5 with Frit 418.

WL	No. of Acceptable EVs	No. of Non-Acceptable EVs	Limiting Constraint(s)
28	1833	2087	Homg
29	2374	1546	Homg
30	2674	1246	Homg
31	2838	1082	Homg
32	3174	746	Homg
33	3560	360	Homg
34	3904	16	Homg
35	3920	0	none
36	3920	0	none
37	3920	0	none
38	3902	18	lowv (11) T_L (7)
39	3722	198	lowv (139) T_L (59)

All of the EVs produced acceptable glasses at WLs of 35 to 37%. Access to lower WLs becomes limited by the homogeneity constraint (which can be relaxed for the sludges without ARP). Access to higher WLs becomes limited by process-related constraints (lowv and T_L) rather than durability-related constraints. As expected, the

---

[a] JMP™, Ver. 6.0.3, SAS Institute Inc., Cary, NC (2005).

WL window within which all of the glasses are acceptable is smaller in the Variation Stage than in the Nominal Stage assessment. However, given the large variation applied to the potential SB5 composition and particularly to the Na_2O concentration, the performance of Frit 418 is excellent. Assuming the homogeneity constraint can be relaxed for both sludge-only and coupled operations, all of the EVs could be processed over a 28-37% WL interval with only 18 of the 3920 EVs failing lowv or T_L at 38% WL.

CONCLUSIONS

SRNL recommended that the DWPF utilize Frit 418 for initial processing of SB5. Nominal and Variation Stage assessments predicted that Frit 418 would form an acceptable glass when combined with SB5 over a range of WLs, typically 30-41% based on the nominal projected SB5 compositions. Frit 418 has a relatively high degree of robustness with regard to variation in the projected SB5 composition, particularly when the Na_2O concentration is varied. The acceptability (chemical durability) and model applicability of the Frit 418–SB5 system were verified experimentally through a variability study.[3]

Frit 418 was not designed to provide an optimal melt rate with SB5, but was recommended for initial processing until better data on the composition of SB5 are received. Experimental testing to optimize a frit composition for a good balance between projected operating windows and melt rate will then be completed.

REFERENCES

[1]Choi, A. S., "Aluminum Dissolution Flowsheet Modeling in Support of SB5 Frit Development," *U.S. Department of Energy Report WSRC-STI-2008-00001, Revision 0,* Washington Savannah River Company, Aiken, SC (2008).

[2]Edwards, T. B., K. G. Brown and R. L. Postles, "SME Acceptability Determination for DWPF Process Control," *U.S. Department of Energy Report WSRC-TR-95-00364, Revision 5,* Washington Savannah River Company, Aiken, SC (2006).

[3]Raszewski, F. C., T. B. Edwards and D. K. Peeler, "Sludge Batch 5 Variability Study with Frit 418," *U.S. Department of Energy Report SRNS-STI-2008-00065, Revision 0,* Savannah River National Laboratory, Aiken, SC (2008).

CERAMIC COATED PARTICLES FOR SAFE OPERATION IN HTRS AND IN LONG-TERM STORAGE

Heinz Nabielek[1,2], Hanno van der Merwe[1], Johannes Fachinger[2], Karl Verfondern[2], Werner von Lensa[2], Bernd Grambow[3], Eva de Visser-Tynova[4]
1) Pebble Bed Modular Reactor (Pty) Ltd, Centurion, South Africa
2) Forschungszentrum Jülich, Germany
3) Ecole de Mines, Nantes, France
4) Nuclear Research & Consultancy Group, Petten, The Netherlands

ABSTRACT

The fully ceramic core of a High-Temperature Gas-Cooled Reactor (HTR) consists of graphite internals and graphitic fuel elements that contain coated particles. These are fissile kernels that are surrounded by several coating layers. In BISO particles, the coating consists of a buffer layer and a dense pyrocarbon layer, in the TRISO particles there is a SiC interlayer interspersed between two dense pyrocarbon layers. While coatings originally had only been introduced for the handling in the fuel, they have been optimized in the meantime to withstand operational gas pressures, to retain fission products generated during irradiation and to keep them inside in accidents. The latest development showed the dense coating layers also to be efficient in long-term storage of HTR fuel.

INTRODUCTION

The coated particle consists of a 500 μm UO_2 kernel, a buffer layer and a series of dense layers: inner pyrocarbon, SiC and outer pyrocarbon; designated a TRISO particle. These layers are mechanically strong and retain fission products throughout irradiation within the limits for HTR typical operation temperatures und burnup.

A variety of coating failure mechanisms has been analyzed and studied in past years to understand and quantify the mechanical and chemical behavior of the compound structure of the coated particles under operational and accidental conditions. The most elementary in-reactor effect is the build-up of gas pressure in the free volume of the porous buffer layer inducing circumferential stresses that ultimately may exceed the tensile strength of the SiC leading to the release of fission products. Major emphasis is therefore placed in the determination of the release of stable and long-lived Xe and Kr as the indicator of a coating failure and the formation of CO. Both strength and strength distribution of SiC and the pyrocarbon layers are the most important material input parameters to enable failure predictions with fuel performance codes.

Recent strong interest in high burn-up fuel for better economy and waste reduction and coated particles that can operate at very high temperatures (e.g., hydrogen production for fuel cells) will need continued modeling work and code development. Scoping calculations with the German code development PANAMA have been useful for exploring the operating range of the fuel up to 20% FIMA.

The potential failure mechanisms including statistics on manufacturing variations, extreme irradiation conditions leading to kernel migration and fission product attack due to accident conditions like core heatup, water ingress, air ingress and others have to be regarded. The coating layers retain their strength and fission product retentiveness in accidents limited to 1600°C. The HTR core is designed such that temperatures are restricted to keep particles intact and retentive.

Direct disposal of spent spherical HTR fuel elements is an attractive option, because fuel and fission products are already well contained. It is the reference concept for the management of the 1 million fuel elements with BISO and TRISO particles from the operation of the two German high temperature reactors AVR and THTR.

The demonstration program investigates the long term safety under final disposal conditions and determines the radionuclide release in case of contact with disposal site relevant aqueous phases.

Several investigations show that the long term safety is related to intact coating layers, especially the SiC layer, and their dissolution behavior of the fuel kernel. The porous graphite matrix may allow access of water that comes into contact with coated particles.

For the long-term storage of spent HTR fuel we need to know, if the containment function of the particle coating layers remain effective over long periods of time. For this purpose, the PANAMA code will be further extended to cope with the additional internal He pressure from alpha decay, the effects of a corrosive environment and external pressures.

Both in normal operation, in accidents and in long-term storage of HTR fuel, we intend to quantify on the protective functions of the coating layer on every single particle rather than large external barriers and containments.

HTR FUEL DESCRIPTION

TRISO coated fuel particles imbedded in graphite form the basis of fuel elements utilized in most high temperature reactor designs. Modern HTR fuel is designed around the following envelope[1]:

- Maximum core temperatures of 1200°C (operation) and 1600°C (accidents);
- Maximum burnup of 12% FIMA, potential for 20% FIMA;
- Maximum fast neutron dose of 6×10^{25} m^{-2} (E>16 fJ);
- Maximum power densities of 12 MW/m^3;

Fuel elements are the first and most important safety feature of HTR designs. Fundamental safety functions are:

- Confinement of radioactive materials and control of operational discharges, as well as limitation of accidental releases;
- Control of reactivity through negative temperature coefficients and core design;
- Removal of heat from the core by core materials and core design.

The fundamental safety functions are achieved by ensuring that high quality fuel is used, operating temperatures and fuel burnup is within fuel specification and chemical attack on the fuel is prevented.

Fuel design criteria are derived from the fundamental safety functions and from the functions that the fuel must perform in the environment in which it will be used. The main criteria can be described as follows:

- Fuel integrity must always be maintained throughout the whole lifecycle of the fuel.
- Fuel must be able to withstand transport and handling stresses during shipping and handling, and remain intact under all expected and design reactor conditions.
- Fuel must contain fission products for the entire lifecycle of the fuel.

To achieve the design criteria, coated particles failure fractions must remain low enough not to cause any significant radiological risk to operating personnel and the general public. It is therefore imperative that chemical attack on fuel and core structures is prevented. Fuel integrity is observed by monitoring gaseous fission product release from the fuel in the primary coolant.

FUEL DESIGNS

All HTR fuel designs can be described as either prismatic or spherical fuel designs. Various prismatic fuel designs have been developed, in particular, in the USA and Japan[2]. These were used in various reactors of which Fort St. Vrain (USA) and HTTR (Japan) are the most significant. The Japanese example is presented in Figure 1.

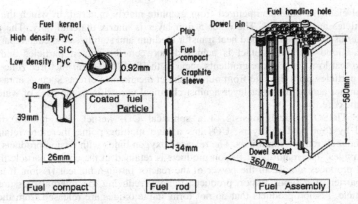

Figure 1. Japanese reference pin-in-block fuel design.

Modern spherical fuel design is based on the German reference fuel for the HTR-MODUL (High Temperature Reactor – Modular)[3] and is presented in Figure 2. This fuel design was extensively tested and evaluated in Germany and the Netherlands. Spherical fuel was tested on a large scale during 21 years of AVR operation at Jülich, and in the 300 MW$_e$ THTR reactor, both in Germany.

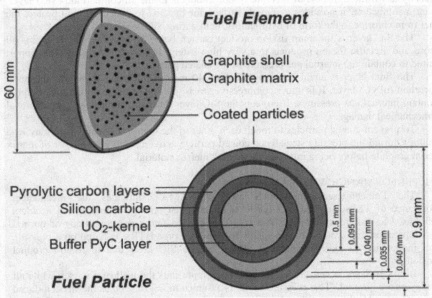

Figure 2. Spherical fuel element and coated particle for pebble-bed reactors.

Fuel elements are manufactured from graphite matrix material, in which the TRISO coated particles are imbedded. The outer 5 mm layer is matrix material only. The graphite matrix material functions as a good heat transfer medium and stabilizes the coated particles in the sphere. Good thermal contact is achieved between the coated particles and matrix material, so that low temperature gradients occur in the fuel. The outer fuel free zone protects the coated particles from damage from outside direct mechanical effects such as abrasion and shock. It further acts as a barrier layer against chemical corrosion in the case of water or air ingress in the core.

The TRISO particle consists of a spherical UO_2 kernel, 500 micron diameter, surrounded by four coating layers. UO_2 has a high melting point, therefore retaining its integrity under all reactor conditions. The released oxygen binds with fission products to form immobile oxides. The majority of fission products is retained in the coated particle this way. The kernel produces almost all the power of the reactor through nuclear fission. It acts as a retention barrier of gaseous fission products, thereby reducing the internal pressure of the coated particle. Fission products that do not form stable oxides are released from the kernel through a diffusion process. All fission products are therefore retained or have their release reduced by the UO_2 kernel.

The kernel is surrounded by a 95 micron low density carbonaceous layer known as the buffer layer. This layer acts as a sacrificial layer, allowing the kernel to swell under irradiation, to stop recoiling fission products and to provide the void volume for fission gases released from the kernel. The rest of the layers are therefore protected from recoiling fission products and excessive internal pressure by the buffer layer.

The next layer is made up of dense pyrocarbon, 40 micron thick and known as the inner pyrocarbon (iPyC) layer. It forms an impenetrable barrier to gaseous fission products, and slows down the transport of metallic fission products to the silicon carbide (SiC) layer. During manufacture, it provides a smooth surface for the SiC to adhere to, and protect the kernel from chlorine in the form of hydrochloric acid during SiC deposition.

The SiC layer is the main fission product barrier, being 35 micron thick. It retains all gaseous and metallic fission products to a very high extent. It provides the structural support required to contain the internal gas pressure in the coated particle.

The final layer is again dense pyrocarbon, 40 micron thick and known as the outer pyrocarbon (oPyC) layer. It is under compressive stress, putting pressure on the SiC, helping to contain internal gas pressures. It protects the SiC layer during manufacture from chemical and mechanical damage.

To prevent coated particles to touch each other in the matrix material which may lead to failures during the pressing stage, each coated particle is overcoated with a layer of matrix material graphite before being mixed with the bulk matrix material.

FUEL PERFORMANCE WITH PANAMA

The Forschungszentrum Jülich code PANAMA[4] simulates the mechanical performance of TRISO coated fuel particles under given normal operation and accident conditions. The failure probability, that is of importance under the conditions of normal reactor operation and core heatup accidents for modular type HTRs, is based on a pressure vessel model and includes degradation effects on the SiC layer due to fission product corrosion.

In the pressure vessel model, the SiC layer represents the wall of a vessel, while all other layers are ignored. This pressure vessel is assumed to fail as soon as the stress induced in the SiC layer by the internal gas pressure exceeds the ultimate tensile strength of silicon carbide. The probability for a pressure vessel failure of a particle is a function of time and temperature and can be described by means of a Weibull relationship.

The tensile strength is a material parameter whose median value and modulus is derived from dedicated experiments like the Bongartz brittle ring test[5]. SiC strength is weakened under irradiation; both median strength and modulus decrease with neutron fluence at irradiation above $1000°C$[6].

The effect of fission product corrosion is transformed into an effective thinning of the SiC layer, representing the pressure vessel wall, at a volume corrosion rate according to Montgomery[7], thus leading to a sooner failure of the coated particle at given conditions.

The internal gas pressure is calculated by applying the gas state law to the generation of fission gases Xe, Kr, and reaction gas CO. The amount depends on the yield of stable fission gases, the burnup, the number of oxygen atoms produced in the kernel, and the temperature and irradiation time.

Oxygen production in the particle kernels as a result of the fissioning of ^{235}U or ^{239}Pu is strongly dependent on the irradiation history and to a great extent on the type of kernel. Corresponding relationships for CO/CO_2 molecules generated were derived from tests at Seibersdorf, Austria, covering an irradiation time up to 550 days and a temperature range between $950-1525°C$[8].

The validation of the PANAMA model has been made against numerous experiments with spherical fuel elements heated at accident temperatures in the range of $1600-2500°C$. Good agreement between predicted particle failure and experimentally derived failure from Kr release measurements had been achieved. For fuel exposed to extreme irradiation conditions, however, the calculated failure fraction has shown the tendency to overpredict experimental failure. Further developments are on the way to provide for the mechanisms that give the coated particle even greater performance potential.

FINAL DISPOSAL BEHAVIOR OF HTR FUEL

The HTR fuel element is a complex multibarrier system with respect to long term radionuclide mobilization under final disposal conditions. Each barrier contributes to the immobilization of the radionuclides in a final repository. To develop a model for long term performance calculations it is necessary to understand the behavior of each component[9]. The fuel kernel of UO_2 or $(Th,U)O_2$ is a confinement matrix for a large fraction of generated fission products and actinides. Since this matrix is stable to dissolution in water, it constitutes a barrier against radionuclide release upon ground water access. The matrix does not confine a certain fraction of radionuclides such as fission gases xenon and krypton, cesium and iodine in the fuel kernel, because they segregate to grain boundaries and buffer porosity.

The fuel kernels are surrounded by dense and water-resistant coating layers that have to be disintegrated before an aqueous phase could come into contact with the fuel kernel. Therefore the coating layers can be regarded as another effective barrier.

The graphite matrix of the fuel element will restrict water access to the coated particles and represents the outer barrier.

The radionuclides in the graphite matrix may come into contact with aqueous phases, if the groundwater penetrates the storage cask. However, the radiotoxic inventory of the graphite matrix is negligibly small in comparison to the inventory in the fuel kernels.

For the fuel kernels, two cases have to be considered: a few fuel kernels with a defective or failed coating and the major number of kernels (> 99 %) with an intact coating. The access of aqueous phases to the failed particles will be controlled by the pore system of the graphite matrix by a diffusion process. Radionuclides in the defective kernels, which were segregated to grain boundaries, will probably be rapidly dissolved in the pore fluid and migrate slowly by diffusion to the surface of the fuel pebble. The major fraction of the radionuclide inventories of these kernels is still retained and they are dissolved only very slowly as determined by the dissolution rate of the fuel kernel matrix.

Most fuel kernels are protected from water access by the coatings, which consist of different layers. The BISO particle has a porous buffer layer and a dense pyrocarbon layer. The TRISO particle has an additional SiC interlayer in between the dense pyrocarbon layers. A release of radionuclides from segregated phases and from the fuel matrix will only occur, when an aqueous medium penetrates these layers.

Each of these barriers has its own properties and behaves differently in aqueous phases. In order to obtain a long-term prediction for behavior during final disposal, it is necessary to investigate the different barrier materials and construct a model for the whole fuel pebble. The main processes are given in Figure 3.

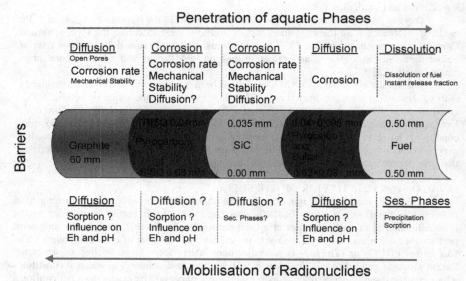

Figure 3. Main processes for radionuclide mobilization from a HTR fuel element.

Leaching tests with irradiated spherical fuel elements show two clear-cut phenomena with respect to retention of ^{137}Cs and [154]Eu (Figure 4):

- Intact particles retain the fission products nearly completely;
- Release from defective/failed particles and from the sphere matrix is fast, but this contribution is small.

These results of leaching experiments are consistent with the hypothesis that mobile radionuclides in the graphite matrix will be released fast and that the main inventory is retained by the intact coated particles.

All the results obtained so far show – as was expected – that the SiC layer represents the most important long term barrier of the coating materials. Therefore, it is necessary to investigate the behavior of this layer under repository conditions by leaching experiments. The chemical resistance of SiC and Zr – both as powders and in the form of coating layers – has been determined for salt domes, granite-based repositories and clay formations. The leaching tests were conducted in oxidizing and reducing atmospheres and at different temperatures. Three kinds of leachant (granite water, clay-pore water and Q-brine) have been used to simulate disposal conditions.

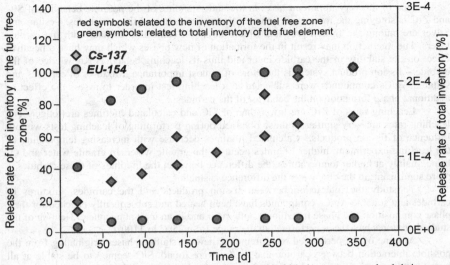

Figure 4. Leaching behavior of an HTR fuel pebble in concentrated salt brine.

A model has been developed to predict material lifetimes based on the experimental leaching tests: uniform leaching rates have been assumed as uniform corrosion rates. On this basis, lifetimes for a 30 µm thick layer are predicted to be between 1000 and 10,000 years. Under reducing conditions, at low temperatures, lower leaching rates for ZrC powder are found than for SiC powder, but at higher temperatures SiC is superior. In any case, the lifetime of the SiC coating guarantees a safe enclosure of the major inventory of fission products that will decay in the first 1000 years (Figure 5). Only the small amount of defective or failed particles will contribute to a release in this time period.

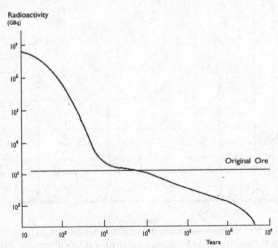

Figure 5. Decay in radioactivity of high level waste
(from OECD NEA 1996: Radioactive Waste Management in Perspective).

One of the very important tasks in the understanding of the possible behavior of SiC and ZrC is studying the interaction between fission products and these carbides. Cesium and Silver are among the fission products which may well come into contact with the carbide layers. The interaction may result in the formation of new phases which may have a negative effect on the stability of the carbide layer and thus its leaching behavior. Knowledge of the carbide – fission product system is therefore of utmost importance. Mixtures of carbides and suitable Ag/Cs compounds were subjected to a leaching test in order to assess the effect of additional phase formation on the behavior of the carbides.

Leaching rates of ZrC are lower than of SiC and calculated lifetimes are longer. The leaching rates are very similar to those obtained during performing of leaching tests without influence of fission products. Calculated lifetimes decrease with increasing temperature of the leaching experiment, higher lifetimes occur in the granite water. In granite water and Q-brine, mainly at higher temperatures, the difference between the lifetimes of these carbides is more significant; in the clay water the difference is smaller.

To study the interaction between fission products and the carbides, mixtures of carbides and suitable Ag/Cs compounds have been heated and subsequently checked for their phase composition and phase transitions, both in air and in an inert atmosphere. Heating of the studied mixtures have been performed in the range from 25°C to 1100°C.

At none of the performed measurements, new crystalline phases originating from the possible interaction between carbide and Ag/Cs were found. SiC seemed to be stable at all tested conditions, ZrC oxidized to ZrO_2 (mixture of cubic, tetragonal and monoclinic phases) at high temperatures, this oxidation took place in the inert atmospheres as well (Figure 6).

The program is continuing to study the stability of coating layers during interaction with fission products.

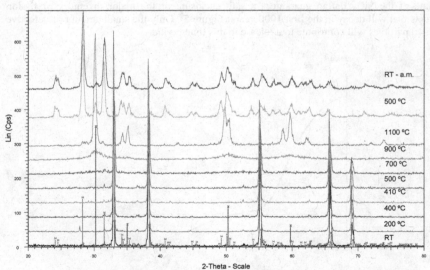

Figure 6. XRD measurement of the ZrC – Cs system in inert atmosphere
[ZrC (red pattern), $CsNO_3$ (orange pattern), ZrO_2 (cubic, blue pattern)
and ZrO_2 (monoclinic, green pattern)].

MODELING LONG-TERM FUEL BEHAVIOR UNDER REPOSITORY CONDITIONS WITH PANAMA

For the prediction of spent HTR fuel performance in a repository, it is necessary to know whether coated particles are able to keep their containment function under long-term storage conditions. An upgraded version of the PANAMA TRISO fuel performance code may be used whereby several effects have to be considered additionally:

- New time scales need to be introduced adjusting the present scale of months and years to tens of thousands of years.
- Spent fuel stored under the typical conditions of a salt mine repository is expected to get (after a certain delay time) into contact with the salt environment leading to a remobilization of fission products from the particle kernel, if the coating is defective or has failed. The corrosive attack from the salt environment on a coated particle will begin at the outside of the outer pyrocarbon layer. Investigations were conducted in Jülich to derive corrosion rates under various conditions for both irradiated and unirradiated material[10].
- An additional delay may be assumed for the slow diffusive transport process of water through the porous matrix material of the fuel element.
- The corrosion mechanism given in the original version of PANAMA, which refers to a corrosive attack by fission products on the silicon carbide layer at 1600-2000°C, is not relevant in a spent fuel repository at temperatures < 200°C.
- Another long-term effect in spent fuel is the radioactive decay of the actinide fission products by alpha-particle (helium) emission. Helium gas will be generated in the particle kernels and may lead over the long disposal times to a significant pressure increase inside the coated particle enhancing the probability of a pressure vessel failure.

An idea of the quantities of helium produced over the disposal time is given in Table I[11]. A more thorough approach could be the evaluation of the detailed calculations of the radioactivity inventory of the existing spent HTGR fuel in Germany[12]. The helium gas would be conservatively assumed to be completely released from the kernel and to fully contribute to the inside pressure load upon the SiC layer.

Table I. Helium Production in Coated Particles during Long-Term Storage.

Time [years]	Helium [10^{18} atoms/g of UO_2]	Helium Pressure [MPa]
40	2.4	0.35
1000	11	0.65
10,000	23	1.3
100,000	41	2.5

Other effects that might have an influence on the long-term storage performance of the fuel particles are the pressure from outside (rock formation above the repository), decay heat production, self-irradiation, or material fatigue mechanisms, e.g., the long-term weakening of the SiC layer strength.

CONCLUSION

Stability and fission product retention has been demonstrated for operating and accident conditions for current HTR designs. Further potential to higher operating temperatures and higher burnup are being explored. Under final disposal conditions, HTR fuel is well packaged: only the radionuclide inventory of defective or failed coated particles can be mobilized. The main part of the radionuclide inventory is sealed by the coating layers. The life time of the SiC coating can be assumed to last 1000 to 10,000 years. Investigations are on-going for the long term interaction between SiC and fission products.

REFERENCES

[1]H. van der Merwe and J. H. Venter, HTR Fuel Design, Qualification and Analyses at PBMR, PHYSOR 2006, Vancouver, Canada (2006).
[2]K. Verfondern (Ed.), Fuel Performance and Fission Product Behavior in Gas Cooled Reactors, IAEA TECDOC 978, International Atomic Energy Agency, Vienna, Austria (1997).
[3]G. H. Lohnert, et al., The Fuel Element of the HTR-Module, a Prerequisite of an Inherently Safe Reactor, *Nucl. Eng. Des.,* **109**, 257-263 (1988).
[4]K. Verfondern and H. Nabielek, PANAMA Ein Rechenprogramm zur Vorhersage des Partikelbruchanteils von TRISO-Partikeln unter Störfallbedingungen. Report Jül-Spez-298, Research Center Jülich, Germany (1985).
[5]K. Bongartz, et al., The Brittle Ring Test: A Method for Measuring Strength and Young's Modulus on Coatings of HTR Fuel Particles, *J. Nuclear Materials,* **62**, 123-137 (1976).
[6]H. J. Allelein, et al., The Behavior of HTR Fuel under Irradiation, 7th Int. Conf. on Structural Mechanics in Reactor Technology, Chicago, USA (1983).
[7]F. Montgomery, Fission Product SiC Reaction in HTGR Fuel, Report GA-905837, General Atomics, San Diego, USA (1981).
[8]E. Proksch, et al., Production of Carbon Monoxide during Burnup of UO_2 Kernelled HTR Fuel Particles, *J. Nuclear Materials,* **107**, 280-285 (1982).
[9]J. Fachinger, M. den Exter, B. Grambow, S. Holgersson, C. Landesman, M. Titov, and T. Podruhzina, Behavior of Spent HTR Fuel Elements in Aquatic Phases of Repository Host Rock Formations. *Nucl.Eng.Des.,* **236**, 543-54 (2006).
[10]T. Podruzhina, Graphite as Radioactive Waste: Corrosion Behavior under Final Repository Conditions and Thermal Treatment, Report Jül-4166, Research Center Jülich (2005).
[11]D. Roudil, et.al., Long Term Behavior Studies of HTR Fuel Particles under Disposal Conditions: Issues, Program and Initial Results, Proc. 3rd Int. Top. Meeting on High Temperature Reactor Technology, HTR-2006, Paper J 0000065, Johannesburg, South Africa (2006).
[12]D. Niephaus, Referenzkonzept zur direkten Endlagerung von abgebrannten HTR-Brennelementen in CASTOR THTR/AVR Transport- und Lagerbehältern, Report Jül-3734, Research Center Jülich, Germany (2000).

Author Index